MODERN RADIO SCIENCE

ICSU Press Symposia

Genetic Manipulation: Impact on Man and Society Cambridge University Press (1984)

Striga: Biology and Control IDRC (1984)

H$^+$-ATPase (ATP Synthase): Structure, Function, Biosynthesis Adriatica Editrice, Bari, Italy (1984)

Progress in Bio-organic Chemistry and Molecular Biology Elsevier Science Publishers (1984)

Global Change Cambridge University Press (1985)

Membranes and Muscle IRL Press (1985)

Integregation and Control of Metabolic Processes: Pure and Applied Aspects Cambridge University Press (1987)

The Cytoskeleton in Differentiation and Development IRL Press (1987)

The Pharmocology of Nicotine IRL Press (1988)

Mechanisms of Vascular Action IRL Press (1988)

Modern Radio Science Oxford University Press (1988)

MODERN
RADIO SCIENCE

Edited by

A. L. CULLEN

Department of Electronic and Electrical Engineering
University College London

Published for the
International Union of Radio Science and the
ICSU Press by
Oxford University Press
1988

Oxford University Press, Walton Street, Oxford OX2 6DP
Oxford New York Toronto
Delhi Bombay Calcutta Madras Karachi
Petaling Jaya Singapore Hong Kong Tokyo
Nairobi Dar es Salaam Cape Town
Melbourne Auckland
and associated companies in
Berlin Ibadan

ICSU Secretariat, 51 Bd de Montmorency, 75016 Paris

Oxford is a trade mark of Oxford University Press

Published in the United States
by Oxford University Press, New York

British Library Cataloguing in Publication Data
Modern radio science.
1. Radio engineering
I. Cullen, A.L.   II. International
Union of Radio Science
621.3841
ISBN 0–19–856223–3

Library of Congress Cataloging-in-Publication Data
Modern radio science / edited by A.L. Cullen.
p.   cm.
Includes bibliographies and index.
1. Radio waves.   2. Ionospheric radio wave propagation.   3. Space
plasmas.   4. Optical fibers.   I. Cullen, A. L. (Alexander Lamb),
1920–   . II. International Union of Radio Science.
III. International Council of Scientific Unions.
QC676.6.M63   1988   537.5'34––dc19   88–5325
ISBN 0–19–856223–3

Printed in Great Britain
at the University Printing House, Oxford
by David Stanford
Printer to the University

Preface

This book arose from the thought that the excellent series of special lectures which have recently been such a valuable feature of the General Assemblies of the International Union of Radio Science should be made available in a more permanent form to a wider section of the scientific community. The lectures, though useful to the specialist, are written in such a way as to be intelligible to the non-specialist, and as such to aid in the cross-fertilization process which URSI regards as so important.

URSI's interests are wider than the word 'radio' in its title suggests, and this is at once evident in the first lecture by Wolinski on laser measurement. The next lecture by Felsen provides an up-to-date review of spectral theory in relation to wave propagation. It is followed by a discussion by Csibi of queuing and coding theory, and this leads into Okoshi's account of coherent optical fibre communications. Three lectures relating to radio-wave propagation follow: the first by Crane deals with relatively high frequencies for which the ionosphere is not normally very important; Rishbeth then gives a concise account of the present state of knowledge of the ionosphere relevant to radio propagation; and last in the group Kurth and Shawhan discuss waves in plasmas, including space plasmas. This leads naturally to Welch's account of the latest developments in radio astronomy. Next, Mme Seguin gives a possible scenario for future communications networks, and we return to optical techniques with Midwinter's lecture on computing and switching. Finally, Axford and Sagdeev describe the exciting observations of Halley's comet.

My hope and belief is that this permanent record of the special lectures will be of value to all who have an interest in radio science, both inside and outside the URSI community.

London   A . L . C .
November 1987

Contents

Contributors

W. I. AXFORD
Max-Planck-Institut für Aeronomie, D–3411 Katlenburg–Lindau, FRG

R. K. CRANE
Thayer School of Engineering, Dartmouth College, Hanover, NH 03755, USA

S. CSIBI
Technical University of Budapest, Budapest, Hungary

LEOPOLD B. FELSEN
Department of Electrical Engineering / Computer Science, Weber Research Institute, Polytechnic University, Farmingdale, NY 11735, USA

W. S. KURTH
Department of Physics and Astronomy, The University of Iowa, Iowa City, IA 52242, USA

J. E. MIDWINTER
Department of Electrical and Electronic Engineering, University College London, Torrington Place, London WC1E 7JE, UK

T. OKOSHI
Department of Electronic Engineering, University of Tokyo, 7–3–1 Hongo, Bunkyo-ku, Tokyo 113, Japan

H. RISHBETH
Rutherford Appleton Laboratory, Chilton OX11 0QX, UK

R. Z. SAGDEEV
Space Research Institute, Academy of Sciences, Moscow 117810, USSR

HELGA SEGUIN
Centre National d'Etudes des Télécommunications, 38–40 rue du Général Leclerc, 92131 Issy-les-Moulineaux, France

S. D. SHAWHAN
NASA Headquarters, Washington, DC 20546, USA

Wм. J. WELCH
Radio Astronomy Laboratory, University of California, Berkeley, CA 94720, USA

W. WOLINSKI
The Institute of Electron Technology, University of Technology, Warsaw, Poland

1

Laser measurement 1968–1987 and beyond

W. WOLINSKI

ABSTRACT

The report, being a general introduction to laser metrology problem, presents examples of laser application to tracing straight lines, to determining angular deviations and the reference plane, in measurements of distance, displacement and distortion, to measurements of atom and particle spectra, and for the construction of length and frequency standards. These examples display not only the achieved level of the measuring methods but also the progress reached during the last 20 years.

INTRODUCTION

In 1968 the URSI Conference on Laser Measurement was held in Warsaw. Even at this time, not so long after the invention of laser radiation, (Maiman 1960), the far-reaching role to be played by the latter in different fields of measurement was fully appreciated. When looking over the proceedings of this Conference (Smolinski, Hahn, 1969) we find many very interesting research results on: the development of optical frequency standards, the application of lasers in interferometry at a large difference of the optical path and of optical radar, laser measurements of plasma parameters and the measurement of element size by means of diffraction picture analysis as well as much work on the properties of lasers themselves. Nevertheless the 20 years which have elapsed since then, raised the level, expanded the range and, in the first place, spread the application of laser metrological equipment.

Now, passing to the survey illustrating progress in this field, I would like to recall briefly the properties of laser radiation from which the possibilities of its application in metrology directly result.

Laser radiation is emitted in a single direction, determined by the optical resonator, the laser beam having a very small divergence angle (of the order of several milliradians) The whole power, emitted within this small solid angle, gives a high luminance

of source (W/m^2sr), higher by some orders of magnitude than that of
sources emitting light in conformity with Lambert's law. For the
fundamental TEM_{ooq} mode, the beam intensity distribution is
displayed by the Gaussian curve. The above mentioned beam, by means
of very simple optic systems, can be collimated still better and
focused (on small surfaces of a diameter of several to several tens
of micrometers) reaching large power and energy densities and thus
evoke nonlinear phenomena, such as light harmonics, autocollimation
and bistability, not known before the invention of lasers.

Laser-radiation is, to a high degree, monochromatic, having a
small spectral width. According to the well known equation of
Schawlow and Townes, the limiting spectral width of the operation of
a laser in a single mode can be determined from the expression:

$$\Delta \nu_L \simeq \frac{2 \pi h \nu \, (\Delta \nu_{pass})^2}{P}$$

where: $h\nu$ -quantum energy, $\Delta h\nu_{pass}$ - half width of the
        resonance curve of passive resonator. P - laser output
        power.

He-Ne lasers: λ = 632.8 nm, ν = $4.73*10^{14}$ Hz, $\Delta \nu_{pass}$ = 0.5 MHz,
P = 1mW, the value of $\Delta \nu_L$ determined from this equation amounts to
$5*10^{-4}$ Hz. Actually different destabilising factors extend this
width by several orders of magnitude. In specially designed and
electronically controlled lasers, a short-term line-width of about
1Hz is reached. Solid lasers, as compared with gas ones, have an
interior monochromaticism.

The laser radiation is coherent in time and space, ie, it shows
a perfect interference capability. The two beams into which a single
beam is divided and which pass two different optical paths interfere
with one another. The difference of these optical paths or the so-
called coherence time L = $c \Delta t$ (where c - light velocity and the
coherence path $\Delta t \simeq 1/\Delta \nu_L$) may reach many kilometers. Similarly,
two beams emitted from different points of the source and having
passed different optical paths will interfere. From among all
classical sources, only low pressure and small current discharge
tubes, emitting spectral line groups of widths resulting from the
Doppler effect, show a time coherence, reduced to the difference of
optical paths of about 80 cm. The above mentioned features as well
as the possibility of choosing lasers operating in the range from
soft X-rays to far infrared (several hundred micrometers) and with
continuous duty or pulse duty even down to femtosecond ranges, the
tunability of the wavelength of some lasers over a wide spectral
range and the possibility to obtain large powers and energies, means
that it is difficult, today, to find a field of science and
technology   where lasers are not applied,

in particular, in metrology. Lasers are applied in the following main fields of metrology:

- geodesic measurements - of straight line, angle deviation, reference plane and distance;
- interference measurements - of quality of transparent media, displacement, distortion of small surfaces, dynamic distortion-holography, seismic disturbance, size of elements by analysing diffraction pictures and surface state etc.
- spectroscopic measurements - by absorption methods or by methods based on nonlinear phenomena and the application of spectroscopic methods for: construction of length and frequency standards, detection and discrimination of chemical compounds (methane, HpD, isotopes), analysis of the surface state and contamination of media, measurement of the flow of gases and liquids and of rotational speed, etc.

Each of the above mentioned fields covers many different procedures and many different designs of the measuring equipment. In the following part of this report, I will limit myself to a brief consideration of topics which, from the aspect of wide application, progress of science and technology, contributed new values and played an essential role.

GEODESIC MEASUREMENTS

The application of lasers in geodesic measurement allowed for:

- carrying out measurements in conditions of limited visibility and at points with difficult accessibility;
- reduction of the duration of measuring procedures accompanied by a raised measuring accuracy;
- automation of geodesic works such as:
    - testing of the distortion of hydrotechnical structures, bridges, viaducts, crane tracks, crane guides, shaft hoists, etc,
    - levelling of longitudinal and transverse profiles, carrying out by means of the tunnel method and in trenches as well as laying of pipelines,
    - multi-storey building, high accuracy alignment of components of large machines,
    - control of the operation of all kinds of working machines (dredgers, bulldozers, etc).
    - standardization of geodesic optical instruments,
    - satellite bearing.

In all the above mentioned works, excepting the last one, laser levelling instruments, theodolites, interscans, plumbers, and rotational equipment, are applied. In these equipments, most often provided with automatic levelling and plumbing, the

measuring axis or plane is determined by a laser radiation beam.
The equipment operates with optoelectronic auxiliary devices
determining precisely the centre of the incident beam (eg,
quadrant detectors placed on levelling staffs) or maintaining
their position within the beam or measuring plane, recording
simultaneously deviations appearing in time.

All the above mentioned equipments are provided with
integrated He-Ne lasers emitting a wavelength λ = 632.8 nm of a
power ranging from 0.5 to 5 mW in the TEM_{ooq} fundamental mode.

When setting the spot position on the staff by eye at a
distance of 100 m a precision from 1.3 to 1.8 mm is obtained for a
beam focused to 12 mm dia. When quadrant detectors are used, this
precision of the determination of the axis rises to 0.2 mm. At
distances up to 20 m with good focussing of the beam and using a
CDD mosaic detector a measuring precision of the position of the
beam axis reaches about 0.05 mm.

The determination of a plane by means of a rotating beam
does not afford good precision. The latter is $\pm$ 20 mm at a
distance of 250 m. This equipment, however, allows for the
simultaneous levelling at many points of an area of this radius
or a simple control of a bulldozer levelling the area.

The determination of the plumb-line (perpendicular) is
carried out by means of a compensator diffracting the horizontal
laser beam through a right angle and maintaining it in this direction
either by the coincidence of the output beam with that reflected
from the mercury surface mirror or by such design of the equipment,
using the gravitation force as to maintain the equipment and the
beam perpendicular. The latter method allows for determining the beam
axis with a precision $\pm$10 mm at a distance of 250 m.

In geodesy engineering practice the necessity arises to
determine the space coordinates of structures, eg, of a tunnel.
Tachymeters are applied for this purpose. After projecting the
laser point upon the structure, the horizontal and the vertical angles
are measured with a precision of 1 angular minute. The distance is
measured by the principle of a variable base of the instrument and a
constant parallactic angle at the aim point. These instruments,
intended for the measurement of small distances up to 60 m, assure
a precision of distance measurement of $\pm$ 1 cm.

The development of geodesive methods and equipment was
directed, not only at raising their accuracy, but at providing
them with microprocessor data-processing systems, the extension
of their measuring capability and the construction of specialised
systems. For instance, the equipment for determining planes,
provided with detectors and a timer measuring the time interval

between pulses obtained from the reflection of the rotating beam
from appropriately set reflectors, also measures simultaneously the
angle with an accuracy up to several angular seconds (the head
turning at a speed of 600 - 60,000 rpm controlled with a quartz
generator). Another specialised equipment controls a tunnel
cutting disk. This equipment not only leads the disk along a
straight line but is also able to deviate it in a programmed
manner from the latter.

Distance measurement from some hundred meters up to some
scores of kilometers (80 km at earth level and pure air) is
carried out using two methods. The first consists in the
modulation of the beam of a continuous duty He-Ne laser (= 632.8nm)
with a frequency 0.1 - 40 MHz and after its reflection either
from the retroreflectors or from the natural object, the precise
measurement of the phase shift 2 . The approximate expressions
for the distance and the measuring error are as follows (Holejko,
1981):

$$L = \frac{\lambda_s}{2}(n + \phi) = \frac{v}{2f_s}(n + \phi)$$

$$\Delta L = \left[\frac{\Delta v}{v} + \frac{\Delta f_s}{f_s}\right] L + \frac{\lambda_s}{2}\Delta\phi$$

where: L = measured distance: λ_s, f_s = length and frequency of
standard wave; n = total number of waves λ_s in 2L; v =
wave velocity.

It can be seen that one term of the error expression depends
upon the distance and the second is constant, depending on the
length of the standard wave and the measuring accuracy of the phase
shift. Carrying out measurements of the same distance at different
frequencies of the standard wave, the absolute value of this
distance can be determined. The manufacturers of this kind of
equipment assess the measuring error either as $\Delta L = \pm 1$ mm or
10^{-6} x L for reflection from a retro reflector or as $L = \pm 3$ mm
or 10^{-4} x L for a diffuse reflection. The error adopted is the
higher value.

The second method is analogous to that used in microwave
radars. The emitted light pulse, after its reflection, returns to
the receiver stopping the timer measuring the time needed for
passing the double path. For this purpose, NdYAG pulsed lasers
with Q-switched optical resonators are used most often. For
distances of the order of 1.5 km, semiconductor lasers are used.
In special equipment, because of the atmospheric window, pulsed

molecular CO_2 lasers emitting a wave of 10.6 m, are applied. As
an example, an anti-tank laser telemeter can have 5 km range and
measuring accuracy of 3 m, where the source consists of a pulsed
CO_2 laser TEA type emitting 60 ns pulses. Two pulses assure the
measurement. Ranges of 1500 m at a visibility of 500 m have
been demonstrated with the use of this equipment.

        Another group comprises equipment intended for geodesive and
geophysical satellite research. By measuring the distances
earth-moon and earth-artificial earth satellites, not only can the
position of the station be determined, but also data on the earth's
rotation, the gravity field, shifts of poles and the dynamics of
oceans can be obtained. The equipment operates on the same
principles as those described above but for the fact that the pulse
is reflected from a set of retroreflectors provided on the satellite
 comprising from ten to several hundred reflecting elements. The
satellites are placed on different orbits of an apogee from 500 to
2000 km. The measuring accuracy is determined by the standard
deviation. For symmetrical pulses this standard deviation of the
distance measurement is determined by the expression (Szydlak 1985).

$$\sigma_R = A \frac{T}{\sqrt{(n\ N)}}$$

where: A = proportionality factor, T = duration of generated
        pulse, n = number of received photons, N = number of
        measurements per second.

        Hence results the requirement for laser emitters used in
this kind of equipment. They are to generate pulses of large
energy, short duration and a (possibly) high recurrence rate. In
order to ensure that a maximum number of photons reach the
satellite, a beam with a minimum deviation angle (10^{-5} rd) is to be
produced. At first ruby-crystal lasers (wavelength 694.3 nm) with
a Q-switched optical resonator producing pulses of energy of about
1J, width 20-30 ns, repeated at a rate of 0.1 Hz were used. In spite
of large diameter emitter and receiver devices, more than 2.5 m,
the accuracy was only 1 m. The second generation of this equipment
used pulsed ruby-crystal lasers of pulse width from 2 to 5 ns. with
a Gaussian radiation distribution and a repetition rate of 1 Hz.
A measuring accuracy, eg of the distance to the satellite LAGFOS
(apogee 5,900 km) of about 10 cm was reached. Equipment of the
third generation used NdYAG lasers emitting waves of length
λ = 1.06 m converted to its 2nd harmonic λ = 530 nm to
which the detectors used are very sensitive. The lasers operate,
mostly, in a mode synchronisation system with selection and
amplification of pulses of width 20 to 200 ps. These lasers may
also be ones with active modulation of the Q-factor and time-
variable losses of the resonator. Beside the considerable
shortening of pulses, their repetition rate can be raised even up to

20 Hz. The main phenomena, rendering measurement difficult and bringing about the necessity of introducing many corrections, are: attenuation, dispersion and refraction of the radiation curving the beam path. Equipments similar to that described above are the so called lidars of range 300 m to 15 km used for measuring the dynamics of clouds, smoke, dust etc.

INTERFERENCE MEASUREMENTS

Within the group of interference measurements a very important role is to be attributed to equipment allowing for the determination of displacements with an accuracy up to a fraction of the light wavelength (Dukes, Gordon, 1970) (Baldwin et al 1971), (Holejko 1981). Two measuring methods are used. One is based on the direct counting of interference striae accompanying the displacement of the mirror constituting one of the arms of a Michelson interferometer (Fig 1). The radiation source used is a He-Ne laser operating at a single frequency maintained with a stability better than 5 MHz per day (frequency instability of about 10^{-8}). To determine the direction of changes, the image of the striae is divided into two beams, with a quarter-wave plate inserted in the path of one of them. The signals obtained from the detectors have phases shifted by 90^0, the discriminator comparing both pulse series brings about the addition or subtraction of the number of pulses displayed on the counter. In most cases this equipment affords an accuracy of $\lambda/4$, ie, for the laser used about $0.15\,\mu$m. An equipment based on the use of the Doppler mirror effect (Fig 2) is also used. The He-Ne laser of frequency instability 10^{-9} generates, due to the Zeeman effect in the transition $3s_2 \Rightarrow 2P_4$, two frequencies f_1 and f_2. In order to facilitate their discrimination, these beams have mutually perpendicular linear polarisation. The reference is provided by the frequency difference $(f_2 - f_1)$ and the signal, taking into account the Doppler effect produced by the mirror moving with a velocity v, is the frequency difference $f_2 - (f_1 \pm \Delta f)$. The difference of these two signals gives the Doppler frequency $\Delta f = \pm 2f_i(v/v_L)$ where v_L = light velocity in the measuring conditions. The displacement can be determined from the following expression:

$$\Delta L = \int_o^T v\ dt = \pm \frac{\lambda}{2} \int_o^T \Delta f\ dt$$

To increase the resolution, resolution expanders multiplying Δf are applied additionally behind the detectors. A measuring resolution of the order $10^{-2}\,\mu$m is obtained for distance up to 60 m. The series connection of expanders allows for raising the resolving power by one order. The accuracy of the displacement measurement is bound by the accuracy of the frequency measurement and amounts to

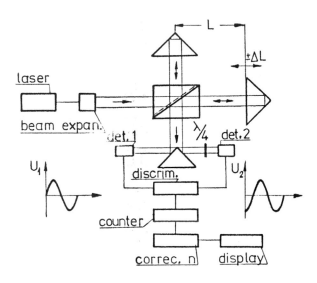

Fig.1. **Layout** of a single-frequency interferometer for measuring displacement.

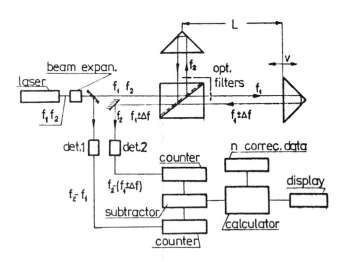

Fig.2. **Layout** of a double-frequency interferometer for measuring displacement.

10^{-7}. These interferometers are controlled automatically by
microprocessors, taking into account corrections due to the
variation of humidity, pressure and temperature of the air.

Equipment used for the precise measurement of displacement
allowing for the automation and increased precision of many
manufacturing processes, play an outstanding role in technology
today. For instance, according to Laser Focus (p 21 March 1986),
equipment for displacement measurements constitutes 57.3% of the
total number of laser equipments sold for microelectronics for a
sum of $1255 million in 1985. A rise in sales from $719 million up
to $1220 million is foreseen in 1990.

Another group of measurements developed since 1978 concerns
holographic interferometry. It allows for touchless nondestructive
quality control of products and measurement of the shape of objects,
measurement of small displacements, deformations and vibrations,
detection of noise sources in equipment and of the paths along
which noise is transmitted. Holographic interferometry if applied
in tests of cool and hot plasma, research on diffusion,
crystallisation, shock-waves, flow patterns, etc. The greatest
advantage of these tests is the fact that they can cover the whole
of a large object, their accuracy being nevertheless of the order of
a fraction of a micrometer. This measuring technique is of
particular importance in the case of extreme reactions, eg, in
rocket, aircraft and motor industries. The measurements are
carried out by taking a hologram of the test object and then
subjecting it to the action of specified forces, a second
exposure is made on the same photographic plate. Interference of
the two images displays all deformations and defects. This method
allows one, moreover, to test the dynamics of stresses by taking a
hologram with two high-power laser pulses quickly following one
another. For this kind of measurement generally two-pulse ruby
lasers (λ =693.3 nm) generating pulses of an energy of the order of
one tenth Joule and length from 5 to 50 ns with a precisely controlled interval between them ranging from 100 to 400 μs are used.
These laser heads require correction of the radiation coherence.
Also NdYAG pulsed lasers with 2nd harmonic transformation (λ = 530 nm)
can be used for this purpose.

How surprising and novel the applications of this technique
may be can be illustrated by the holographic interferogram of the
vibration of the human throat and face when singing (Fig 3)
(Pawluczyk 1980). It can be used for the assessment of the
professional ability of vocalists. Today holographic interferograms
are used for qualitative tests. Nevertheless the technical
literature provides some information on the computer processing of
holographic interferograms, furnishing data on the mechanical
and operational features of the objects tested.

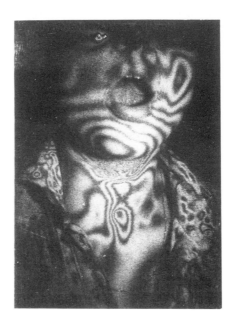

Fig.3. Holographic interferogram of vibrations of the
surface of throat and face.

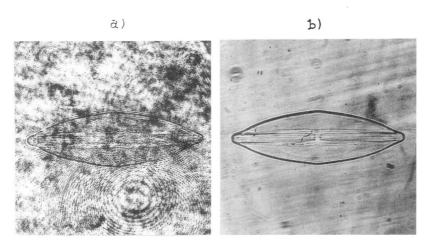

Fig.4. Holographic pictures of a diatom (bacillario-
phycea) recorded by microscopy : a) without attenuation of
coherent noise. b) with attenuation of coherent noise

One of the main phenomena limiting resolution power and simulation contrast is the unwanted diffraction and interference effect, known as coherent noise, appearing in the laser light. An interesting very efficient method for the elimination of this noise is suggested in the reports (Mroz et al 1971) (Pawluczyk Mroz 1973). It consists in changing the direction of incidence of the beam lighting up the object by means of a rotating glass cube whose axis of rotation is perpendicular to the beam axis. Above a certain velocity of rotation of this cube, the grain images shifted along the interference striae, and whose sources lie outside the plane of the image, are blurred. This is the so-called single-direction time-averaging method of coherent noise. This method has been successfully applied in the development of holographical and holographic-interference microscopes. Fig 4 represents two holograms of a biological preparation without and with attenuation of the coherent noise (Pluta, 1980).

SPECTROSCOPIC MEASUREMENTS

Spectroscopy is a field whose development is to be particularly attributed to the application of lasers, mainly of tuned lasers among which the first place is occupied by dye lasers ($\lambda = 0.3 \ldots 1 \mu$m) of continuous and pulsed duty. Besides the latter, tuned excimer lasers ($\lambda = 0.12 \ldots 0.32 \mu$m), lasers with coloured centres ($\lambda = 0.7 \ldots 3 \mu$m), Raman spin-flip type lasers ($\lambda = 4 \ldots 5.5 \mu$m and $\lambda = 8 \ldots 12 \mu$m), solid-state laser ($\lambda = 0.5 \ldots 30 \mu$m), parametric oscillators ($\lambda = 0.5 \ldots 4 \mu$m) and nonlinear frequency mixing systems ($\lambda = 35$ nm $\ldots 100 \mu$m) are also used. The application of lasers to the well-known absorption methods has permitted simplification of the measuring systems and also raised the resolution and sensitivity of methods. However, a qualitative jump in this field is due to nonlinear spectroscopy characterised, in the first place, by the elimination of Doppler line broadening. Saturation spectroscopy (Letochow, Czebotajew, 1982), (Gawlik 1985) is the basic method. Fig 5 provides a diagram of the measuring system. Two beams, the saturating one and the sampling one of frequency ω are propagated in the test medium in opposite directions. An atom moving with a velocity component v_z along the propagation direction of the radiation sees a beam of frequency $\omega(1-v_z/c)$ due to the Doppler effect. The laser beam reacts selectively with atoms of such a v_z that a resonance occurs between the atomic transition frequency ω_0 and the beam frequency in the moving system, ie, $\omega(1-v_z/c) = \omega_0$. Hence, $v_z = v_{res} = c(\omega - \omega_0)/\omega_0$ and on the curve of the population of the upper level, a peak appears at the beam's v_z. The sampling beam of a low radiation intensity, not changing the distribution of the population of levels $N(v_z)$ and reacting with the group of atoms of $(-v_{res})$, undergoes attenuation after passing through the medium. The measured radiation intensity is a measure of the absorption factor of the medium. When the laser is tuned strictly to resonate with the

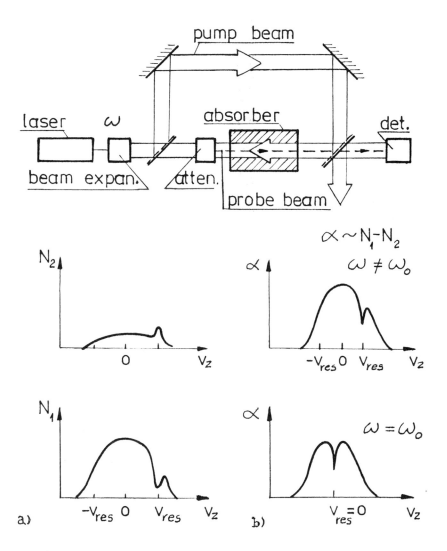

Fig.5. Layout of equipment for saturation spectroscopy:
a) population of lower (N_1) and upper (N_2) energy levels
of atoms versus their velocity component conform with the
direction of radiation propagation v , under reaction of a
strong laser beam of ω close to $ω_o$. b) absorption factor
versus v_z for ω close to $ω_o$ and ω = $ω_o$.

tested transition $\omega = \omega_0$ and when $v_{res} = 0$ both beams react with the same group of atoms. The feeble beam samples the medium saturated already by the strong beam and is less attenuated after passing then beyond resonance, displaying a narrow peak. Doppler broadening is eliminated by the selection of a narrow velocity class.

During the last decade polarisation spectroscopy has been developed, which is a certain kind of saturation spectroscopy. The difference consists in the fact that the sampling beam detects not the change of the saturated absorption factor but the change of the dispersion properties of the medium caused by the saturating beam (Wieman, Hansch, 1976) (Gawlik, Series, 1979). This allows for the elimination of the Doppler-broadened background and only narrow lines are recorded. The amplitude of the signal can be increased by beating it with the sampling beam. This brings about, at the same time, a change from the Lorentz shape to a dispersion one which can be used, for instance, as a signal for very precise stabilisation of the laser. The progress achieved in measuring resolution is illustrated by the curves reported (Hansch et al .. 1979) for a line structure α, a wavelength $\lambda = 656$ nm, Balmer's series of a hydrogen atom. The resolution power of these spectra is so high that it displays the hyperfine structure of lines, allowing for such precise measurement of Rydberg's constant ($R = 2\pi^2 me^4/h^2$) that it is one of the most accurately determined physical constants known ($R = 10973731.521(11)$ m^{-1}), (Amin et al 1981).

Besides the above mentioned measurements, many other spectroscopic methods raising the resolution and sensitivity and bringing about the possibility of investigating selected atoms or levels have been developed. These methods comprise: two-photon spectroscopy, velocity-selective optical pumping spectroscopy (V.S.O.P.), fast ion beam spectroscopy, labelling spectroscopy, spectroscopy with ultrasonic molecular beam, etc. I want only to call your attention to the two topics which will influence the future development in this field. The first is the application as detectors of lines of photodiodes or CDD matrices allowing for a spacial reproduction of the whole spectrum image. Matrices are often coupled with opto-electric image converters and amplifiers of the microchannel plate type, allowing one to obtain a signal in each channel of a sensitivity corresponding to the detection of single photons. This renders possible a multichannel reproduction of the spectrum which is essential for the investigation of dynamic processes so short that, until now, they were impossible to record. The second topic concerns the study in the range of subnatural spectroscopy reaching beyond the natural line width resulting from spontaneous emission energy levels. From among the methods quoted in the literature I would like to call attention to the method using nonstationary optical pumping effects (Gawlik et al ... 1982)

(Gawlik, 1986). Firstly, the narrow structures could be attributed
to Zeeman coherences which are induced by the probe beam revealed
by differential light shifts that remove the degeneracy of the
magnetic sublevels. Secondly, nonstationary optical pumping into a
level that cannot be excited further by any of two laser beams.
This method allows us to investigate atoms and molecules in the
gaseous phase. Some of these molecules, being subject to optical
pumping, ie, those having a suitable structure and a long enough
lifetime of their lower energy levels. Measurements are made by
means of equipment similar to that used in polarisation spectroscopy
but provided with additional adjustable diaphragms and attenuators.
The difference of these measurements as compared with the
polarisation method consists in the fact that the probe beam has a
high radiation intensity and, similarly to the pumping beam, strongly
disturbs the medium. Fig 6 represents test results for the line
D_1 of Na atoms. Curve (a) is obtained by the polarisation method.
Curve (b) represents a fragment of the spectrum obtained using the
new method, representing the hyperfine components $3S_{1/2}(F=2) \rightarrow$
$3P_{1/2}(F=1$ and $F=2)$ with an additional resonance of cross-over type
in the centre between them. An indentation of 2.6 MHz half-wave
width and much narrower than the natural width of the line D_1
amounting to 10 MHz, is visible in the centre of one of the component
lines. The simplicity of the system together with its subnatural
accuracy and high sensitivity constitutes the undoubted advantage
of the method.

        Spectroscopy is responsible for important achievements in the
field of fundamental research, such as:

        - Statement of maintained parity in atomic spectra;
          verification of the theory of the unification of reactions
          (Forston, Cets 1980) (Stacey, 1984);
        - direct observation of quantum transition (Aspect et al, 1982)
          (Perrie et al ... 1985);
        - determination of new value of the electron radius (Dehmelt
          1985);
        - observation of the modification of spontaneous emission
          through a resonant cavity (Gabrielse, Dehmelt, 1985)
          (Hulet et al .. 1985);
        - observation of new states of the electromagnetic field of
          minimum fluctuations, known as squeezed states (Wu et al ..
          1986);

and many many others. Also the application of spectroscopy in
different fields is very wide, for instance, for;

        - analysis and diagnostics of media and surfaces;
        - verification and control of chemical reactions
          (photochemistry, photomedicine, separation of isotopes)
        - ultra-sensitive detection of radiation, mainly of microwave

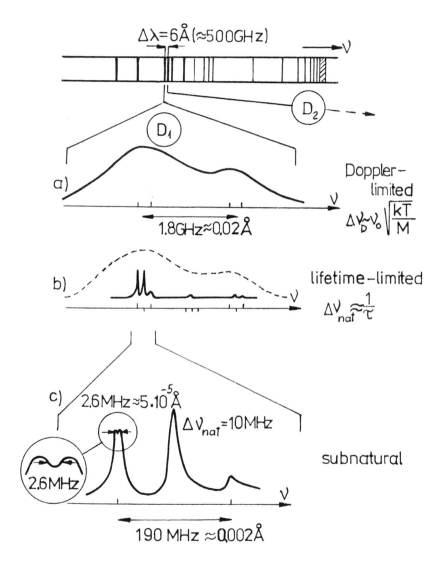

Fig.6. Set of measurements of line D_1 sodium Na:
a) spectral profile of line D_1 with account taken of the hypersubtle structure and the Doppler broadening, b) the "Doppler-less" method of polarisation spectroscopy, c) fragment of the spectrum (b) recorded by new method.

type (by means of Rydberg atoms);
- construction of gyroscopes and flowmeters;
- construction of new length and time standards.

Let us look closer at the last group of applications mentioned.

Gas lasers, such as He-Ne and CO_2 ones, because of the small spectral width of their radiation ($\Delta\nu$), are much better as length standards than the standard established in 1960 in the form of a Kr^{86} discharge tube emitting for the transition $2p_{10} \geqslant 5d_5$, a wavelength 605.7802105 nm. Moreover, the coherent laser radiation, easily treated by heterodyne procedures, allows for realisation of a frequency standard. A laser suitably stabilised by means of external negative feedback systems may, at the same time, be used as a length and frequency standard. Dip Lamba stabilised He-Ne lasers, or lasers using for stabilisation the Zeeman splitting effect of the line λ =632.8 nm, assure an instability of frequency of the order 5×10^{-8} and serve as length standards in interference measurements. A quantum standard comparable with the caesium (Cs^{133}) standard (frequency instability $(1 ..2) \times 10^{-13}$, ν = 9192631770 Hz) have been developed stabilising the laser radiation frequency by means of the saturable absorption phenomenon. Reports on this topic were presented at the URSI conference of 1968. At this time a frequency instability of 10^{-9} ... 10^{-11} was obtained, ie, from four to two orders of magnitude worse than that reached today. An atomic or molecular standard in the form of an absorption cell placed, generally, inside the laser resonator must fulfil several conditions. The absorption line of the standard must coincide with the laser emission line and the transition frequency must be constant and reproducible at best with an accuracy specified for a quantum standard. The latter condition is fulfilled by many different transitions between undisturbed electric and magnetic fields and collisions, energy levels of quantum systems. Moreover, taking into account the possibility of a servosystem used for stabilisation with reference to the centre of the resonance curve, the relative width of the resonance curve of the standard must not exceed the desired magnitude of the frequency instability more than $10^3..10^4$ times. At 10^{-13} this relative width should be of the order $10^{-9} ..10^{-10}$, ie, several orders of magnitude narrower than brought about by Doppler broadening ($10^{-5} .. 10^{-6}$). A resonance minimum appears in the case of an absorption cell placed inside the laser resonator when the generated frequency is tuned to the frequency of the absorption curve centre. In consequence, a resonance maximum used for stabilisation is in the middle of the laser output power curve plotted versus frequency. A certain frequency shift occurs between the maximum power peak and the middle of the absorption curve due to the phenomenon of frequency pulling. This is the method used to stabilise a He-Ne laser generating a radiation of wavelength λ = 3.39 μm and provided with a CH_4 absorption cell.

In the spectral line P7 of the oscillatory-rotating band ν_3 of methane having a lifetime of about 0.01 s, a narrow resonance minimum is obtained. The peak width on the laser output power curve amounts to 300 KHz. Using synchronous detection, a discrimination curve of very steep slope and high signal to noise ratio is obtained. Application of a conventional stabilisation loop keeps the laser frequency within the peak centre (Fig 7). A frequency instability of 1×10^{-13} for averaging times from several seconds and longer and a reproducibility of 10^{-11} have been obtained in this system (Barger, Hall, 1969). Because the wavelength lies in the range of visible light, the application of a He-Ne laser $\lambda = 632.8$ nm is attractive. In this case iodine J^{127} or J^{129} absorption cells are used. The iodine lines coincide precisely with the centre of the Doppler curve of the line $\lambda = 632.8$ nm of the laser, requiring more complex stabilisation systems consisting in the synchronous detection of the third derivative of the auxiliary modulation signal. In this way, a quantum standard of frequency instability from 3×10^{-12} to 1×10^{-11} in an averaging time 1 ... 1000 s and a reproductibility one order worse, ie, 4×10^{-11} to 1×10^{-10} (Quenelle, Wuerz, 1983) is obtained.

For the stabilisation of the operating frequency of lasers, there is also a method according to which the absorption cell is placed outside the laser resonator. This is typical for saturation spectroscopy. A strong beam saturates the medium and the feeble probe beam is reflected from the mirror and propagated in the opposite direction. This kind of equipment needs a high laser output power and high absorption by the cell. However, it has the advantage that the frequency pulling phenomenon does not appear. A typical equipment comprises a CO_2 laser with a SF_6 cell. In this case, a power peak having a width close to the half-width of the natural absorption line is obtained. The above equipment enables a frequency instability of 10^{-12} (Ouhayoun, Borde, 1977) to be reached.

The above brief survey of laser applications, selected from the three main groups of laser applications in metrology, shows that the laser has become an indispensable tool in all fields of technology and science. It not only facilitates and increases the accuracy of methods using well-known principles, but gives access to quite novel possibilities. Metrology is developing along new lines and now applies new methods exerting an essential influence on the worldwide level of science and technology.

ACKNOWLEDGEMENTS

Last but not least I would like to thank Mr W Gawlik as well as Mr S Pachuta and Mr R Nowicki for the materials sent to me which helped, to a large extent, in the preparation of this report.

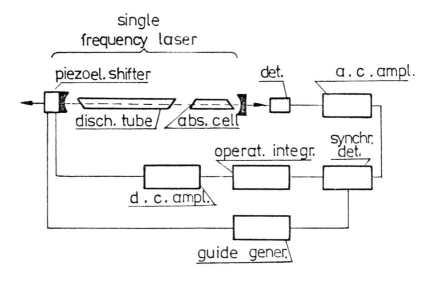

Fig.7. Frequency stabilisation system of laser with internal absorption cell.

REFERENCES

Amin, S.R., Caldwell, C.D., Lichten, W., (1981). Crossed-beam spectroscopy: A new value for the Rydberg constant. Phys. Rev. Lett. 47, 1234-1238.

Aspect, A., Grangier, P., Roger, G. (1981). Experimental test of realistic local theories via Bell's theorem. Phys. Rev. Lett. 47, 460-463.

Aspect, A., Dalibard, J., Roger, G., (1982). Experimental test of Bell's inequalities using time-varying analyzers. Phys. Rev. Lett. 49, 1804-1807.

Baldwin, R.R., Gordon, G.B., Rude, A.F. (1971). Remote laser interferometry. Hewlett-Packard Journal. 12, 14-20.

Berger, R.L., Hall, J.I. (1969). Pressure shift and broadening of methane line at 3.39 m studied by laser saturated molecular absorption. Phys. Rev. Lett. 22, No 1, 4-8.

Dehmelt, H. (1985). Single atomic particle at rest in free space: new value for electron radius. Ann. Phys. Fr. 10, 777-795.

Dukes, J.N., Gordon, G.B. (1970). A two hundred foot yardstick with graduation every microinch. Hewlett-Packard Journal, 8, 2-8.

Forston, E.N., Wilets, L. (1980). Adv. At. Molec. Phys. 16, 319 -

Gabrielse, G., Dehmelt, H. (1985). Observation of inhibited spontaneous emission. Phys. Rev. Lett. 55, 67-70.

Gawlik, W. (1985). Laser spectroscopy. In I-st Symposium of Laser Technology (eds W. Woliński et al), 267-286, ZGPW, Warsaw (in Polish).

Hansch, W., Schawlow, A.L., Series, G.W. (1979). The spectrum of atomic hydrogen. Scientific American 240, 72-86.

Holejko, K. (1981). Precision Electronic Distance and Angle Measuring Instruments. 43, 168-171. Technical Publishing House, Warsaw (in Polish).

Hulet, R.G., Hilfer, E.S., Kleppner, D. (1985). Inhibited spontaneous emission by a Rydberg atom. Phys. Rev. Lett. 55, 2137-2140.

Letochow, W.S., Czebotajew, W.P. (1982). Nonlinear Laser Spectroscopy 26-39. Technical Publishing House, Warsaw. (in Polish).

Mroz, E., Pawluczyk, R., Pluta, M. (1971). A method for coherent optical noise elimination in optical systems with laser illumination. Optica Applicata. 1, No 2, 9-15.

Ouhayoun, O., Borde, C.J. (1977). Frequency stabilization of CO_2 lasers through saturated absorption in SF_6. Metrology 13, 149-150.

Pawluczyk, R., Mroz, E. (1973) Unidirectional optical coherent noise elimination in optical systems with laser illumination. Optica Acta 20, 379-386.

Pawluczyk, R., (1980). Holographic interferometry of opaque object. In Optical Holography (ed. M. Pluta) 205. Polish Scientific Publishers, Warsaw (in Polish).

Pierrie, W., Duncan, A.J., Beyer, H.J., Kleinpoppen, H. (1985) Polarisation correlation of the two photons emitted by metastable atomic deuterium: a test of Bell's inequality. Phys. Rev. Lett. 54, 1790-1793, 2647.

Pluta, M (1980). Holographic microscopy and interferometry. In Optical Holography (ed. M Pluta) 515. Polish Scientific Publishers, Warsaw (in Polish).

Quenelle, R.C., Wuerz, L.J. (1983). A new microcomputer-controlled laser dimensional measurement anlaysis system. Hewlett-Packard Journal 4, 3-7.

Smolinski, A., Hahn, S (inv. eds.) (1969) Proceedings of the URSI Conference on Laser Measurement. Electron Technology (ed B. Paszkowski) 2, 2/3. Polish Scientific Publishers, Warsaw.

Stacey, D.N. (1984). Experiments on parity non-conservation in atoms. Acta. Phys. Polonica A66, 377-397.

Szydlak, J. (1985). Application of lasers to satellite ranging. In 1st Symposium of Laser Technology (eds W. Wolinski et al) 287-308, ZGPW, Warsaw (in Polish).

Wieman, C., Hansch, T.W. (1976). Doppler-free laser polarization spectroscopy. Phys. Rev. Lett. 36, 1170-1173.

Wu, L.A., Kimble, H.J., Hall, J.L., Wu, H. (1986). Generation of squeezed states by parametric down conversion. Phys. Rev. Lett. 57, 2520-2523.

2

Waves and spectra: a modern perspective

LEOPOLD B. FELSEN

ABSTRACT

Electromagnetic and other types of waves are being utilized for communication, detection, remote sensing, etc., in increasingly complicated environments, over wide frequency bands, pushing toward higher frequencies. Spectral methods applied to the spatial and temporal behavior of the field play an important role in the study of wave phenomena under these conditions. Appealing to the local nature of high frequency wave propagation, with the consequent possibility of accommodating propagation and diffraction phenomena of substantial complexity, spectral theories are under investigation that seek to construct and transport optimally contracted wave spectra without compromising essential wave processes. These wave spectra describe physical observables that need to be combined so as to furnish phenomenologically stable parametrizations of complicated wave events. The spectral building blocks in this game are plane waves, rays, beams and, for layered or ducting media, locally adapted (adiabatic or intrinsic) modes, either by themselves or in self-consistent hybrid combination with other wave types. Recent reviews by the writer have dealt with these problems in some detail. Emphasis in the presentation here is on spectral stripping and restoration techniques, and on some pitfalls when the restoration leaves spectral holes.

I. INTRODUCTION

Tracking of high frequency wave fields through complicated propagation and scattering environments is assuming increased importance in applications such as communication, detection and remote sensing. Because of the usually large dimensions of the problem on the scale of the local wavelength, analytically and numerically tractable solution schemes require shrinkage of the problem size by localization. For analytical modeling, spectral analysis and synthesis in the time and space domains has furnished a powerful tool that is formally rigorous, and therefore <u>globally</u> applicable, for special (usually coordinate separable) environmental models. At high frequencies, global spatial wave spectra can be localized around their constructive interference maxima or other critical local wave characteristics generated by the problem geometry. This feature can be employed to accommodate wave phenomena in configurations that depart weakly (over the scale of a local wave-length) from globally separable prototypes; by matching a global prototype locally to each small region on a nonseparable heterogeneous environment, one may employ in that region the corresponding global spectrum, with the recognition that its <u>important</u> contribution is <u>localized</u> around the interference maxima or other critical spectral regions. The spectral shrinkage (or windowing) is problem dependent and must be done with care. The game may be played by various rules (Felsen, 1985), the validity of which can be tested by comparison with corresponding results derived from relevant rigorous prototypes. Clearly, the more drastic the shrinkage, the simpler the resulting wave object. An optimally collapsed spectrum is localized at a single point in the spectral continuum, and the resulting "bare bones" skeletal wave object has no cognizance of other important wave phenomena that may occur. This point spectrum defines a "local plane wave" whose propagation trajectory is a ray defined according to geometrical optics or, more generally, the geometrical theory of diffraction (GTD) (Hansen, 1981). In transition regions of the observational domain, the point spectra (i.e., the rays) of two or more of these skeletal wave fields approach one another, and tracking their smooth interaction requires a spectral umbrella that covers them collectively; thus, by a process usually referred to as <u>uniformization</u>, more spectral flesh must be retained on, or added to, appropriate locations on the spectral skeleton. Starting from a fully fleshed prototype, one arrives at this goal by the selective removal already mentioned. Starting from bare bones, one arrives at the goal by selective spectral restoration. Clearly, the second procedure, essentially an inverse problem of finding the proper spectral continuum from given point spectral data, requires greater sophistification. Local tracking of spectrally contracted observables thus involves alternations between optimally shrunk and minimally fleshed out spectral forms as the propagation trajectory passes from regular into, and out of, transitional domains.

   The scenario sketched above has provided a major focus for modern methods of analysis of space and time dependent wave propagation in generally heterogeneous environments, under conditions that emphasize high frequency effects. Much attention is being given not only to the theoretical basis for spectral stripping and reconstruction, but also to rendering these constructs suitable for numerical evaluation. By spectral localization, problems are parametrized in terms of physical observables because each localized contribution describes a particular wave phenomenon. What spectral representation to choose among various options, and how to combine them most effectively in hybrid schemes, has been the subject of recent reviews by the writer, to which the interested reader may refer (Felsen, 1984, 1985, 1986). These reviews also contain many references which are not included here. In the present discussion, emphasis is placed on the spectral stripping and restoration game, and how applying it inadequately can leave crucial spectral voids, with corresponding numerical algorithms that stabilize on the wrong solution.

II. SPECTRAL UNIFORMIZATION

A. Bulk Waves - Uniformization can follow two alternative routes. By the first, the direct route, the transitional phenomena are modeled in an equivalent global (usually coordinate separable and rigorously solvable) environment adapted to the actual local conditions. This permits use of global spectra and thereby the construction of the full spectral continuum in integral form. High frequency uniform asymptotics applied to these integrals then identifies the pertinent canonical forms incorporating confluences of critical points (stationary phase points and (or) singularities). In this manner, one deduces spectral transition integrals that are expressible as Airy functions near smooth caustics, Fresnel functions near shadow boundaries due to edges, Fock integrals near shadow boundaries due to convex shapes, Weber parabolic cylinder functions near critical angles, etc. (Hansen, 1981). These globally defined transition functions are then patched onto an actual configuration that is simulated locally by the global canonical model. Being uniform, these functions reduce to the isolated and simpler wave events as the observer departs from the transition region. In this direct and conventional approach, the transitional spectra are modeled at the outset, and the simplified regular spectra are deduced therefrom by subsequent reduction permitted when the critical spectral domains separate.

More recent efforts have been directed at pursuing an inverse route. Here, one begins with the simplest localized spectral objects in _regular_ regions, the ray fields, and seeks to construct therefrom the necessary transitional spectra as the observer along the ray path approaches a transition region. This spectral enhancement amounts to putting the proper amount of flesh onto the ray field bare bones spectral skeleton, a much more intricate procedure than removing substance from the originally fleshed out skeletal frame, as in the direct approach. The spectral restoration near caustics, as formalized in the Maslov technique (Maslov, 1972, Chapman and Drummond, 1982, Ziolkowski and Deschamps, 1984), transforms skeletal ray fields into a spectral continuum by a transition to a configuration-spectrum phase space. In this spectral reconstruction, the detailed behavior of an isolated ray field amplitude, which diverges as the observer approaches a caustic, provides the basis for the generation of the proper non-singular object in phase space that negotiates the singularity.

Actually, it is possible to proceed more directly and avoid the rather massive mathematical structure of the Maslov route. This follows from the recognition that the desired uniformization is to be embedded within a plane wave spectral continuum bundle, whose phase behavior varies around the phase along the ray path, and whose spectral amplitudes are to be such as to reduce, by asymptotic evaluation of the spectral integral outside the transition region, to the ray field in question. It has been shown that this more direct reconstruction can be applied not only to caustics but also to other transitional effects like those encountered at shadow boundaries due to edges (Arnold, 1982, 1986), critical angle effects and other diffraction phenomena, as well as to uniformization of trapped local mode propagation through cutoff in tapered layered media (Arnold and Felsen, 1987). In a homogeneous medium, the spectrum is comprised of true plane waves, while in an inhomogeneous medium, the spectrum involves WKB-type local plane waves.

To illustrate what is implied in the former case, express a wave-field $u(\underset{\sim}{r})$ at a point $\underset{\sim}{r}$ as a spectral representation over plane waves $\exp(ik\hat{\underset{\sim}{\xi}}\cdot\underset{\sim}{r})$ as follows

$$u(\underset{\sim}{r}) = \int \tilde{u}(\underset{\sim}{\xi})e^{ik\hat{\underset{\sim}{\xi}}\cdot\underset{\sim}{r}}d\underset{\sim}{\xi} , \quad |\hat{\underset{\sim}{\xi}}| = 1 \tag{1}$$

where $k = \omega/v$, ω being the frequency and v the wave propagation speed in the medium, while $\underset{\sim}{\xi}$ is a two-dimensional spectral variable corresponding to the two-dimensional $\underset{\sim}{\rho}$ domain in configuration space, $\xi_\ell(\underset{\sim}{\xi}) = \hat{\underset{\sim}{\xi}} - \underset{\sim}{\xi}$ and $r_\ell = \underset{\sim}{r} - \underset{\sim}{\rho}$ are the longitudinal components perpendicular to $\underset{\sim}{\xi}$ and $\underset{\sim}{\rho}$, respectively, and $\tilde{u}(\underset{\sim}{\xi})$ is a spectral coefficient. The spectral variable here is the normalized spatial wavenumber $\underset{\sim}{\xi}$, but it can also be propagation angle, etc., and the spectrum should include propagating as well as evanescent waves. The spectral coefficient is assumed to have the form

$$\tilde{u}(\underset{\sim}{\xi}) = A(\underset{\sim}{\xi})\exp[ikS(\underset{\sim}{\xi})] \tag{2}$$

where the amplitude A is slowly varying compared to the strongly oscillatory function $\exp(ikS)$. The composite phase $\psi = (S + \hat{\underset{\sim}{\xi}}\cdot\underset{\sim}{r})$ has stationary points $\underset{\sim}{\xi}_s$ where $(\partial\psi/\partial\underset{\sim}{\xi}) = 0$. Stationary phase evaluation of (1), with (2), yields (Felsen, 1973)

$$u(\underset{\sim}{r},\underset{\sim}{\xi}_s) \sim \frac{C}{k}\left|\det(\partial^2\psi/\partial\xi_1\partial\xi_2)\right|_{\underset{\sim}{\xi}_s}^{-1/2}A(\underset{\sim}{\xi}_s)\exp[ik\psi(\underset{\sim}{\xi}_s)] \tag{3}$$

where C is a collection of constants and $\xi_{1,2}$ denote the one-dimensional components of $\underset{\sim}{\xi}$. Introducing a reference coordinate $\underset{\sim}{r}_o$, and a path length s from $\underset{\sim}{r}_o$, along the straight line trajectory specified by $\hat{\underset{\sim}{\xi}}_s$, one may write

$$\underset{\sim}{r} = \underset{\sim}{r}_o + \hat{\underset{\sim}{\xi}}_s s \tag{3a}$$

with $S \equiv S_o = $ constant at $\underset{\sim}{r}_o$. Thus, $S(\underset{\sim}{\xi}_s) = S_o - \hat{\underset{\sim}{\xi}}_s\cdot\underset{\sim}{r}_o$.

Now, a conventional <u>ray</u> field has the local plane wave form

$$\hat{u}(\underset{\sim}{r},\underset{\sim}{\xi}_s) = \hat{A}(\underset{\sim}{r},\underset{\sim}{\xi}_s)\exp(iks) \tag{4}$$

where the spectral parameter ξ_s is the ray parameter specifying the direction of the ray path on the initial surface r_0, and s is the path length specified in (3a). The local plane wave amplitude $\hat{A}$ varies inversely with the square root of the ray tube cross section, which in turn can be related to the wavefront curvature. The field in (4) can be constructed entirely by ray tracing, following the rules of GTD (Hansen,1981), without recourse to spectral analysis. Equating $\hat{A}(r,\xi_s)$ with the amplitude function in (3) multiplying exp(ikψ), one can determine $A(\xi_s)$, which turns out to be independent of r because the change of $\hat{A}$ with r is compensated by that of $|\det(\ )|^{1/2}$. Thus, $A(\xi_s)$ remains finite even when $A(r,\xi_s)$ diverges, as on a ray caustic where $|\det(\ )|$ vanishes. With the already determined form of $S(\xi_s)$, and replacing ξ_s by the arbitrary spectral variable ξ, the spectral integrand (1) has now been constructed from the independently obtained skeletal ξ_s spectral information in (4). This uniformized spectral object properly negotiates the transitional domain.

B. Guided and Surface Waves – In layered media, the full spectral integrals are constructed initially for the laterally homogeneous coordinate separable prototype, the two-dimensional plane wave spectral variable ξ being associated with the lateral spatial variable ρ parallel to the layer boundaries. In the z-domain perpendicular to ρ, these spectral plane waves experience multiple reflections in a geometric progression, which can be summed into closed form, thus accounting for their collective behavior (Felsen, 1984). The spectral amplitudes resulting from the collective treatment exhibit singularities (resonances) at the discrete eigenvalues ξ_p of the laterally guided modes:

$$1-R_u(\xi_p)R_D(\xi_p) = 0, \quad p = 0,1,2.. \tag{5}$$

where seen from a common reference plane z = constant, R_U and R_D are relection coefficients accounting, respectively, for the upper and lower boundaries of the layer. Reduction of the spectral integral in terms of the residues at these pole singularities furnishes the global normal mode expansion.

When weak lateral variations are included, local enforcement of the spectral resonance condition yields laterally dependent eigenvalues $\xi_p(\rho)$. Merely replacing previously constant depth parameters --for example, the layer spacing h-- by h(ρ) in the laterally homogeneous spectral integral does not alter these spectra sufficiently to generate from the laterally dependent resonances the properly normalized and symmetrized local (adiabatic) mode expansion. To repair this deficiency, it is necessary to apply the lateral spectral adaptation to the entire spectral continuum, thereby changing ξ into $\xi(\rho)$. The spectral adaptation (scaling) is implemented according to the adiabatic invariant

$$R_U[\xi(\rho),\rho]R_D[\xi(\rho)\rho] = \text{const.} \tag{6}$$

which generalizes to arbitrary spectral values the adaptability exhibited by the ξ_p point spectra of the adiabatic modes. The adiabatic mode condition (5) is recovered from (6) by setting the constant equal to unity.

An initially well confined local mode propagating toward the narrowing portion in a tapered layer may experience cutoff in a transition region wherein its behavior changes continuously from the trapped to the leaky (radiative) regime. The bare bones adiabatic mode form associated with $\xi_p(\rho)$ is inadequate to describe the transition phenomena which combine the lateral wave (critical angle) as well as leaky wave effects that are operative here (Kamel and Felsen, 1983; Arnold and Felsen, 1983); in the spectrum, these effects are associated, respectively, with a branch point singularity and a pole. The properly scaled spectral continuum can negotiate passage through the cutoff region and thereby represents the uniformized (spectrally fleshed out) version of an adiabatic mode; the corresponding spectral object has been called an "intrinsic mode". By an independent and more incisive construction (Arnold and Felsen, 1983), the intrinsic mode spectrum can be put together by plane wave superposition, wherein the effect of the previous modal spectral pole is replaced by stationary phase constructive interference. Simple stationary phase evaluation, when applicable, reduces the intrinsic mode integral to the adiabatic mode form. This occurs far from cutoff, on the trapping side, where the stationary phase point in the integrand is well separated from the lateral wave branch point.

To construct the uniformized fleshed out intrinsic mode spectrum from the bare bones adiabatic mode spectrum, it is suggestive, in analogy with what was done for ray fields, to perform the fleshing out with plane waves, but paired so as to account for the standing waves generated by boundary reflections (Arnold and Felsen, 1987). Thus, the single plane waves $\phi = \exp(ik\underset{\sim}{\xi}\cdot r)$ in (1) are replaced by the superposition $\bar{\phi} = (\phi_+ + B\phi_-)$, where $\tilde{\phi}_+$ and ϕ_- distinguish upgoing and downgoing waves, and B is a spectral amplitude attuned to the vertical boundary reflection conditions at any lateral range ρ. The spectral variable $\underset{\sim}{\xi}$ may be based on any convenient parametrization. Again decomposing the spectral amplitude $\tilde{u}(\xi)$ as in (2), one may perform a stationary phase evaluation of the resulting modified integral (1), the composite phase being comprised of S in (2) and the rapidly varying lateral phase in $\bar{\phi}$. The result corresponding to (3) is then matched to the adiabatic mode field, the guided mode analog of the ray field in (4), which can be found directly from normal mode theory applied locally, without need of spectral representations (Weinberg and Burridge, 1974). As in the sequel to (4), this yields the unknown spectral amplitude and phase.

III.   COMPLEX SPECTRA AND GAUSSIAN BEAMS

In what has been discussed so far, general wavefields in homogeneous or inhomogeneous media have been expressed as, and generated
by, a continuous spectral superposition of true or local plane waves,
respectively. By an alternative approach, one may regard wavefields
as being generated by a continuous spatial superposition of localized radiating sources. These two formulations can be considered as
one another's dual because a plane wave is spatially wide but
spectrally narrow whereas for a localized source, the opposite is
true. It is of interest to explore wavefunctions that provide a
continuous transition from one of these elementary building blocks
to the other. Very convenient among these is a Gaussian beam (GB);
by adjusting its waist from very large to very small values, one
passes from the plane wave to the localized source limits (Fig. 1a).
Gaussian beams have received much attention because they simulate
focused wave phenomena encountered in practice (for example, outputs
from lasers or from large reflector antennas); more recently, they
have also generated much interest in respect to their suitability as
basis elements for general wavefield representations. Particularly
attractive is the fact that "bare bones" asymptotic forms generated
by ray fields, for example, which may become singular in transition
regions, can be regularized when these nonuniform field constituents
are replaced by Gaussian beams (Fig. 2). This feature is attributable to the smoothly truncated configurational and spectral extent
of these beams.

A Gaussian beam field in real coordinate space may be generated
by a compact source located in a complex coordinate space (Fig. 1(b))
(Felsen, 1986). It is implied thereby that the spatial wavenumber
spectra synthesizing such a field are complex (evanescent waves), as
are the propagation paths, the complex rays, along which these
spectra propagate. It can now be understood why the transitional
failures of real-source generated ray fields near caustics, shadow
boundaries, etc., are eliminated when such sources are replaced by
complex-source Gaussian beams: the transition regions occur in
complex space, away from the real-coordinate physical domain.
Gaussians as basis elements also have attractive computational
features because both their spatial and spectral extent is windowed
by a smoothly decaying (Gaussian) envelope. This property has been
exploited in what has become known as the Gaussian beam method (GBM)
for representing fields generated by real source distributions as
distributions over Gaussian beams (Cerveny, 1985a).

One approach to evaluating the radiated field due to an initial source distribution is to discretize the integration over the initial surface into patches, and to propagate the fields generated at each patch by ray theory to the observer. If the source distribution or the medium properties are such that focusing takes place, this bare bones ray based propagation algorithm fails at the resulting caustic (Fig. 2(a)). Although this deficiency can be removed by the spectral uniformization procedures described in Section II, this is inconvenient, especially for complicated source arrangements or environments, because the caustics must be located before the damage can be repaired. However, if the ray tube fields were to be replaced by paraxial Gaussian beams, which are propagated with an algorithm only slightly more complicated than that for rays (Cerveny, 1985b), the resulting superposition will be regular everywhere (Fig. 2(b)), and the need for caustic search and transitional corrections eliminated. This concept is the basis of GBM. Within the context of what has been said before, the GB substitution can be regarded as an alternative way of fleshing out the bare bones paraxial ray spectrum. Just what spectral flesh is restored in this manner has been the basic uncertainty besetting GBM: because Gaussian beams have a greater number of assignable parameters than ray fields, the arbitrariness in the choice of beam parameters has prevented GBM from having a priori predictability (Felsen, 1984).

An example (Lu, et al., 1987) serves to illustrate how "wrong" choices of beam parameters generate solutions that stabilize to wrong values of the radiated field. An oscillating line current is located at the point S in a homogeneous half space $z > 0$ with refractive index n_1, which is separated from an exterior homogeneous half space $z < 0$ with refractive index $n_2 < n_1$ (see Fig. 3); the condition $n_2 < n_1$ implies that total reflection into the upper half space takes place when the incidence angle θ of the local plane wave (ray) field, as measured from the z-axis, exceeds the critical angle $\theta_c = \sin^{-1}(n_2/n_1)$. The line source is assumed to be located far enough from the interface so that this surface is illuminated only by propagating (nonevanescent) incident fields which arrive along a cylindrical wavefront (i.e. a set of straight rays) centered at S. The reflected field in $z > 0$ is to be observed at points $P(x,z)$ along the dashed line from S, parallel to the interface.

The reflected field for all x contains the conventional geometrical optics component, but because of the occurrence of critical reflection, it also contains a diffracted lateral wave contribution for $x > x_c$, where x_c identifies the intersection of the reflected ray with the dotted line when the incident ray angle is θ_c (Fig. 3). The lateral wave is launched by the critically incident ray, travels tangent to the boundary on the exterior side, and sheds energy back into the source medium by refraction. For $x < x_c$, the conventional reflected ray total field behaves monotonically whereas for $x > x_c$, the conventional reflected ray and the lateral ray interfere, thereby producing oscillations. The smooth connection between $x < x_c$ and $x > x_c$ ranges takes place in a transition region $x \approx x_c$ that surrounds the critically reflected ray; neither isolated (bare bones) ray theory nor isolated (bare bones) lateral wave theory is applicable there, and a uniformzed fleshed out spectrum is required. This can be constructed from the bare bones spectra by the techniques described earlier (Arnold and Felsen, 1987) which yield, in fact, the well known <u>exact</u> spectrum in this case (Felsen and Marcuvitz, 1973). The resulting integral can therefore be computed <u>globally</u> to provide a reference solution. This is to be compared with what is obtained when the incident field is expressed as a superposition of Gaussian beams, which are then reflected from the interface.

By the recipe of GBM, each Gaussian beam is approximated paraxially (see Fig. 1(a)), and it is reflected from the boundary without attention to the critical angle effect. Because of the smoothing introduced by the GB substitution, the GB field indeed remains finite through the critical angle transition. However, the <u>value</u> of the field depends strongly on the beam stack parameters. Results are shown in Fig. 4 (Lu, Felsen and Ruan, 1987). The precise parameters in the calculation are irrelevant for the illustration here and are therefore omitted; only qualitative designations are given. It is seen that densely stacked very wide-waisted (large b) beams produce excellent agreement with the reference solution whereas sparsely stacked narrow-waisted beams produce poor agreement, with successive improvement for choices in between. It is to be emphasized that the stacking criteria, whether dense or sparse, are those that produce a <u>stable</u> solution; for stability, many wide beams are required to fill the appropriate spectral interval around each ray whereas a few suffice when the beams are narrow.

The behavior noted above can be attributed to the paraxial approximation which strips away portions of the spectrum of the full GB field. It turns out that wide-waisted paraxial beams in a stable stack fill the important spectral domain adequately while narrow-waisted paraxial beams leave spectral voids that cannot be filled even if extra beams are added to what is already a stable stack (Lu, Felsen and Ruan, 1987). Similar conclusions have been reached by White, et al. (1987). This particular behavior of different paraxial beam stacks is peculiar to the critical reflection phenomenon, and it cannot be extrapolated to other diffraction phenomena which have their own peculiar spectral characteristics. This is the dilemma of the paraxially approximated GBM.

The difficulties disappear when each beam field in a stable stack
is computed <u>exactly</u>, without paraxial spectral stripping. Thus, it
is found that all of the options in Fig.4, ranging from densely
stacked wide to sparsely stacked narrow beams, produce complete
agreement with the reference solution. Although numerical computation
of the fully fleshed out complex beam spectra is more difficult than
that of the paraxial spectra, the example illustrates the validity
of GBM if the full beam spectra are retained. This observation is
important because the full spectrum approach may lead to a better
understanding of how beam spectra may be trimmed systematically instead of by trial and error.

IV. SUMMARY

Localized spectral representations of high frequency fields play
an important role in analytical modeling of wave propagation and diffraction in complicated environments. Tracking a wavefield through
successive encounters requires monitoring of the corresponding spectra that describe it, with the goal of shrinking the spectral window
as much as possible without losing those features that are essential.
Optimal shrinkage represents isolated bare bones wave objects parametrized by a point spectrum: a more fleshed out spectral continuum
is required to negotiate transition regions where two or more bare
bones wave objects coalesce. The bare bones objects in bulk media
are ray fields, and in layered media they are local (adiabatic) modes
which express collectively the effect of multiple reflected rays.
Typical transition regions for the former occur around ray caustics,
shadow boundaries, critical reflection boundaries, etc., and for the
latter around cutoff points in guiding regions that cause an initially trapped local mode to become radiative. Spectral stripping,
uniformization and restoration can be effected by various techniques
which undergo continuing refinement. They include asymptotics for
spectral integrals, adiabatic transforms (Hazak et al., 1983; Lu, et
al., 1987), boundary layer and multiscale expansions, phase space
techniques, etc. (for a summary, see Felsen, 1984; Arnold, 1986).
A sample has been included in this presentation.

Especially intriguing because of its potential utility for numerical implementation is the Gaussian beam method. Some of its pitfalls, attributable to the truncated spectral content of paraxially
approximated beams, have been highlighted here. Finding new rules
that combine predictive accuracy with numerical tractability is an
active area of current research. Using properly fleshed out complex
beam spectra provides one approach, but at the expense of numerical
complexity. Rigorous discretization schemes with Gaussian basis
functions in a (configuration-spectrum) phase space lattice
(Bastiaans, 1980; Einziger, et al., 1986) provides another. If
these approaches, and others not yet on the horizon, succeed in making GBM more systematic, hopefully without unmanageable increase in
numerical complexity, one may look forward to quantitative treatment
of successively more complicated propagation and scattering models
encompassing large dimensions at high frequencies.

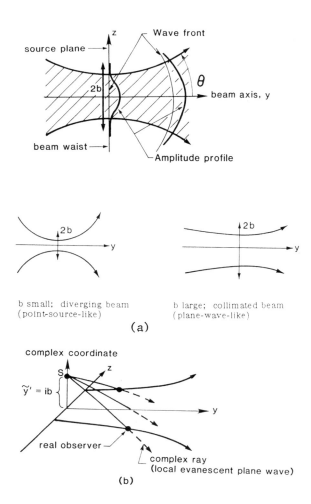

Fig. 1. Two-dimensional Gaussian beam. (a) Beam field in real
space. The beam parameter b̂ is related to the beam width L at
the waist and the far-zone angular width θ as follows:
$L = (b\lambda/\pi)^{1/2}$, $\theta = (\lambda/\pi b)^{1/2}$. "Paraxial" refers to the behavior
near the beam axis. (b) Construction of beam field by real-
space intersection of complex rays emitted by a line source at
the complex location $\tilde{y}' = ib$, $b > 0$.

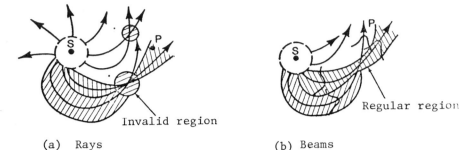

(a) Rays   (b) Beams

Fig. 2 - Paraxial ray and beam tracing.

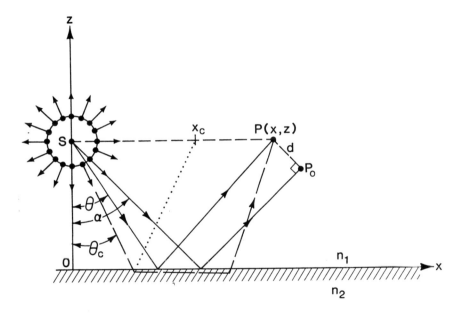

Fig. 3 - Physical configuration pertaining to test case. S = line
source location; P = observer location. Reference (central) rays
emanate from discrete locations (heavy dots) inside patches on
the equivalent source surface surrounding S.   Reflected ray:
solid; lateral ray: long dashes; α = departure angle of central
ray in source surface patch, θ = departure angle of paraxial ray
to P; d = paraxial displacement from central ray. For GBM, the
central rays represent the beam axes. For exactly represented
beams, θ, α and S are complex (see Fig. 1(b)).

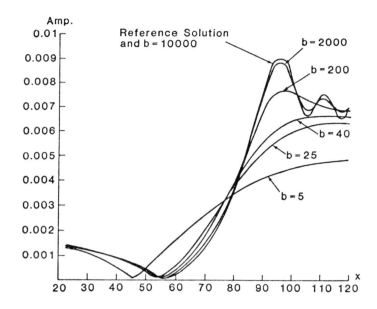

Fig. 4 - Reflected field magnitude along line SP in Fig. 3.
Reference solution and paraxial beam stack solutions for various
beam width parameters b. b = 10,000: wide-waisted; b = 5: narrow-
waisted. Wide-waisted beams must be densely stacked, while
narrow-waisted beams can be sparsely stacked, for stable result.
When the full complex spectra are retained, all options agree
with the reference data (Lu, Felsen and Ruan, 1987).

V.   ACKNOWLEDGEMENT

The research summarized here has been supported by the Office of
Naval Research under Contract No. N-00014-83-K-00214, by the U.S.
Army Research Office under Contract No. DAAG-29-85-K-0180, and by
the Joint Services Electronics Program under Contract No. F49620-85-
C-0078.

REFERENCES

Arnold, J.M., (1982). Oscillatory Integral Theory for Uniform
    Representation of Wave Functions, Radio Sci., 17, 1181-1191.

Arnold, J.M., and Felsen, L.B., (1983).Rays and Local Modes in a
    Wedge Shaped Ocean, J. Acoust. Soc. Am., 73, 1105-1119.

Arnold, J.M., (1986). Spectral Synthesis of Uniform Wavefunctions.
    Wave Motion, 8,135-150.

Arnold, J.M., and Felsen, L.B., (1987). Spectral Reconstruction of

Uniform Wave Fields from Skeletal Nonuniform Ray or Mode Forms, submitted to J. Acoust. Soc. Am.

Arnold, J.M. (1986), Geometrical Theories of Wave Propagation: A Contemporary Review, IEE Proc., 133, Part J, 165-188.

Bastiaans, M.J., (1980). The Expansion of an Optical Signal Into a Discrete Set of Gaussian Beams, Optik, 57, 95-102.

Cerveny, V., (1985(a)). Gaussian beam synthetic seismograms. J. Geophys., 58, 44-72.

Cerveny, V., (1985(b)). Ray Synthetic Seismograms for Complex Two-dimensional and Three Dimensional Structures, J. Geophys. 58, 2-26.

Chapman, C.H. and Drummond, R., (1982). Body-waves in Inhomogeneous Media Using Maslov Asymptotic Theory. Bull. Seismol. Soc. Am., 72, 277-317.

Deschamps, G.A. (1971). Gaussian Beams as a Bundle of Complex Rays, Electron. Lett., 7, 684-685.

Einziger, P.D., Raz, S., and Shapira, M. (1986). Gabor Representation and Aperture Theory, J. Opt. Soc. Am., 3, 508, 522.

Felsen, L.B., and Marcuvitz, N., (1973). Radiation and Scattering of Waves, Prentice Hall, New Jersey. Chapter 4.

Felsen, L.B. (1976). Complex-Source-Point Solutions of the Field Equations and Their Relation to the Propagation and Scattering of Gaussian Beams, Symposia Matematica, Istituto Nazionale di Alta Matematica, Vol. XVIII, Acad. Press, London and New York, 40-56.

Felsen, L.B. (1984). Geometrical Theory of Diffraction, Evanescent Waves, Complex Rays and Gaussian Beams, Geophys. J. Roy. Astron. Soc, 79, 77-88,

Felsen, L.B., (1984). Progressing and Oscillatory Waves for Hybrid Synthesis of Source Excited Propagation and Diffraction, IEEE Trans. on Antennas and Propagation, AP-32, 775-796.

Felsen, L.B., (1985). Novel Ways for Tracking Rays, J. Opt. Soc. Am. A2, 954-963.

Felsen, L.B., (1986). Real Spectra, Complex Spectra, Compact Spectra, J. Opt. Soc. Am. A3, 486-496.

Hansen, R.C., (1981). Geometric Theory of Diffraction, IEEE Press. Distributors: John Wiley and Sons, New York, p. 405. Contains a collection of basic papers.

Hazak, G., Bernstein, I.B. and Smith, T.M., (1983). Integral Representations for Geometric Optics Solutions. Phys. Fluids, 26,684-688.

Kamel, A. and Felsen, L.B., (1983). Spectral Theory of Sound Propagation in an Ocean Channel with Weakly Sloping Bottom. J. Acoust. Soc. Am., 73,1120-1130.

Lu, I.T., Felsen, L.B., and Ruan, Y.Z., (1987). Spectral Aspects of the Gaussian Beam Method: Reflection from a Homogeneous Half Space, Geophys. J. Roy. Astron. Soc., 89, 915-932.

Lu, I.T. and Felsen, L.B. (1987). Adiabatic Transforms for Spectral Analysis and Synthesis of Weakly Range Dependent Shallow Ocean Green's Functions , J. Acoust. Soc. Am., 81, 897-911.

Maslov, V.P., (1972). Perturbation Theory and Asymptotic Method, Dunod, Paris.

White, B.S., Norris, A., Bayliss, A., and Burridge, R., (1987). Some Remarks on the Gaussian Beam Method, Geophys. J. Roy. Astron. Soc., to appear.

Weinberg, H. and Burridge, R., (1974). Horizontal Ray Theory for Ocean Acoustics, J. Acoust. Soc. Am., 55,63-79.

Ziolkowski, R.W. and Deschamps, G.A. (1984). Asymptotic Evaluation of High Frequency Fields Near a Caustic: an Introduction to Maslov's Method, Radio Sci., 19, 1001-1025.

3

Queuing and coding in multi-user communications: ideas, techniques, theory

S. CSIBI

ABSTRACT

A tutorial account is given mainly of cooperative queuing and to some extent, also of multi-user coding approaches of fundamental interest to multiple-access communications. Principles of accessing arbitrarily located users either to each other or to an outside network via a common medium are considered. Delay tolerant users, e.g., individuals accessing to the network via personal computers, as well as delay sensitive users, e.g., talkers and (for broadband media) also people involved in video services, are assumed. The common medium may be a terrestrial radio system, a satellite relay system, some metallic bus, or some metallic or fiber optic loop. An insight is given into the techniques of estimating the most relevant performance characteristics and into some of the underlying disciplines of possible interest also outside of the scope of the present exposition.

INTRODUCTION

Public Telephony has grown during its remarkable hundred years history to a worldwide service with a high number of telephone sets per citizen, and also per km^2, in many countries. However, it has been achieved only since about a decade, and only within some particularly advanced infrastructures, that even mobile subscribers can access to the Public Telephone Network from almost any place within an urban or rural area. In this case we say that a geographically free access to the network has been established (Fig.1)

At present a lot of people, operating teletypewriters or their more recent successors, as well as many users of public data services access to data networks through circuit switched subscriber lines. However this way of local data access is, in many cases, too awkward and too expen-

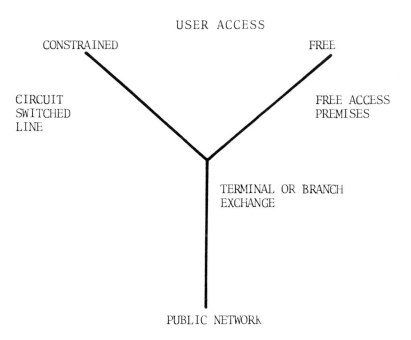

USER ACCESS

CONSTRAINED　　　　　　　　　　　　　FREE

CIRCUIT　　　　　　　　　　　　　　FREE ACCESS
SWITCHED　　　　　　　　　　　　　PREMISES
LINE

TERMINAL OR BRANCH
EXCHANGE

PUBLIC NETWORK

Fig.1. Alternatives for accessing telephone users to a network.

sive, due to the very bursty behavior of most data users.

This is just one of the reasons why data packets communications became so popular within the past fifteen years via metallic buses, metallic and fiber optic loops, terrestrial radio and satellite relay systems for local as well as regional services.

Packet communications offer, since about a decade, flexible and cost efficient means also for a joint access of telephone and data users to local area networks. Accessing jointly data, speech and video users in a similarly flexible manner to broadband local, metropolitan and satellite networks appears, more recently, also a very promosing perspective (Green 1984, Green and Godard 1986).

Much of all these interesting trends is inherently related to radio technologies: as a matter of fact only by these means is geographically free access from any location within an area precisely possible. It appears, therefore,

particularly timely to devote a tutorial presentation to
the methodologies underlying these trends also within the
Scientific Programme at this General Assembly.

OBJECTIVES, REQUIREMENTS AND CONSTRAINTS

Obviously, for achieving free access to networks through
a common medium, in the aforementioned geographic sense,
additional investments are necessary; still the services
involved should be offered for a reasonable cost per user.
Accordingly, the maximum number of users served simultaneously, as well as the complexity of the implementation
are of particular interest.

The size of the simultaneously admissible user population should, of course, be increased only up to a limit at
which the performance of the services provided is still
satisfactory.

The fluency of human and man-machine dialogues pose
one of the most fundamental requirements in this respect;
Viz., the response time (i.e., the overall loop delay
between the parties responding to each other) should be
strictly kept within a tolerable limit.

The obvious difference between the maximum tolerable
loop delay in human and (keyboard and display based)
man-machine dialogs is of particular interest (Table 1).

TABLE 1. Typical loop delay tolerances

	Still tolerable loop delay	
	verbally	in ms
Human dialogue	less than one or two syllables durations	160 - 600
Keyboard and display based dialog	less than usual keyboard operation times	250 - 1000

As far as Public Telephony is concerned, the 0.48-0.56 seconds propagation loop delay of geostationary satellites, is most critical (Fig.2).

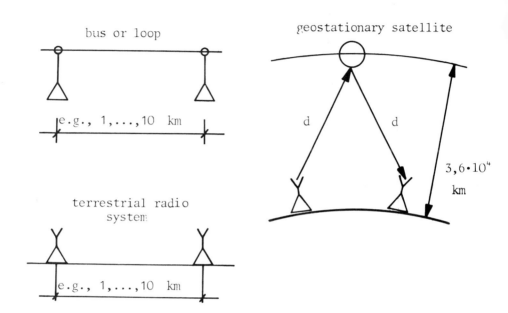

Fig.2. Propagation loop delay: $d_p = 4d$.
(480 ms < d_p < 560 ms)

Errors in the packet flow can, of course, never be entirely avoided, particularly in terrestrial radio and satellite relay systems.

While even rarely occuring errors may cause disturbing clicks in telephony and disappointing interferences on video screens, these two delay sensitive services are usually much more tolerant toward packet errors (and also toward packet losses) as delay tolerant data services.

As a matter of fact the requirements are different and the physical constraints are of distinct significance for these three kinds of services.

COOPERATIVE QUEUING: PRINCIPLES

We use the term cooperative queuing, in the present context, under the following circumstances: (i) several users have to be accessed to a common server (e.g., a common medium) through individual waiting lines, (ii) the access is controlled locally at the place of each user, (iii) the local controllers are just partially informed about the state of the server and the performance of the service, (iv) the local controllers do their best to promote all services simultaneously offered by the system.

Obviously, the principles of centrally scheduled time and frequency division multiple-access schemes are outside of the scope of cooperative queuing.

When automobile drivers meet at a crossing with no traffic lights, and also when people are sitting around a table and talking to each other, all aforementioned conditions, except (iii), are met.

Turning to multiple access technologies, sequentially scheduled loops and random access schemes are two typical examples of cooperative queuing.

In sequential scheduling, used almost exclusively for fiber optic loops (and also frequently for metallic loops) two packet streams are continuously circulating in opposite directions along each member of a double loop. Packets of various length can be either inserted into the packet stream or read and deleted at user interfaces, called nodes, intersecting both member loops at any place, assigned optionally for user access (Fig. 3, Lazar et al. 1985, Lazar and and White 1984.)

Though there is no central control, in this case, for the local insertions and drops, there is still a possibility to watch the performance of the message flow, at any node, before accomplishing these manipulations. The system designer of sequential scheduling should, of course, be interested in acting at each node in a way that appears locally as most favorable, concerning all users involved. Thus we have got a cooperative queuing task in this case.

The common feature of random access schemes is that a packet can be transmitted to the common medium at any time, admitted by the access rule; a copy of this rule should, of course, be available at the place of each individual waiting line. The local control is, also in this case, just partially aware of what other users are doing, and

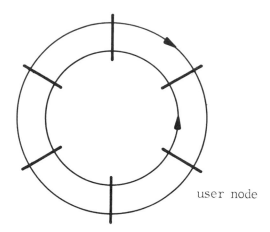

Fig.3. A sequentially scheduled access scheme using
a double loop

what actually is happening in the common medium.

There are random access schemes that can prevent, at
least for a negligible propagation and sensing delay, any
overlap (i.e., collision) of the packets. Others admit collisions, but try to handle these just under light traffic,
by some very simple algorithm. Still others attempt to
handle collisions even under heavy traffic efficiently, by
admitting more complexity.

Carrier sensing schemes (adopted, e.g., in the widespread ETHERNET system) are of the first kind, the ALOHA
schemes (that actually initiated most of the contemporary
work on random access techniques) are of the second kind,
and more complex schemes relying upon particularly efficient algorithms are of the third kind of the aforementioned random access techniques. The initiating of a
packet is admitted either at any time or just at nodes of
some temporary slotting (Fig. 4).

For carrier sensing itself no feedback from the opposite party is needed, just a listening into the common
medium before any transmission. This is done by admitting
an additional function at the place of each user in addition to the packet communication procedure: viz., sensing

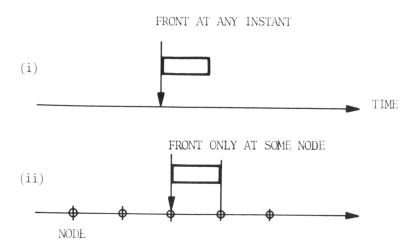

Fig.4. Temporary (i) unslotted and (ii) slotted schemes

the signal present at the considered interface of the
common medium. This sensing function can also be extended
to include collision detection. In addition the main packet
procedure can also be furthered by adopting binary feedback in order to inform the transmiting party about a successful reception at the place of the opposite party
(Kleinrock and Tobagi 1975, Kleinrock 1985, Molle and
Kleinrock 1985, Dallos 1978).

A throughput (i.e. a temporary utilization of the common medium) close to unity is achieved in this way, provided the propagation delay is really negligible; which is
the case for local area networks and local terrestrial
radio services. However carrier sensing is not sensible
for satellite relaying, as the one way propogation delay may
exceed the packet delay in this case.

Pure ALOHA schemes do not rely on any listening into
the common medium, however a binary feedback is needed in
this case, for informing the transmitter about the successful transmission of the most recently launched packet.
If no positive acknowledgement of this arrives within some
waiting period, fixed in advance, another copy of the previously launched packet is transmitted for another trial,
next to a random delay. Only a throughput much less than
unity is achieved in this case. However, this fact does
not much matter provided only a flexible access of just a
couple of users (and not an efficient use of the common

medium) is the objective. (Abramson 1970 and 1985, Tsybakov 1985, Gallager 1985, Saadawi and Ephremides 1984.)

For the third kind of random access schemes there are still at least two choices left for the designer: adopting either temporary blockwise or a temporary free access. (Fig.5. Tsybakov 1985, Gallager 1985, Tsybakov and Mikhailov 1978, Capetanakis 1979a and 1979b, Massey 1981, Mathys and Flayolet 1985.)

In the temporary blocked case any newly arriving demand is accessed to the common medium only as all previously collided demands were met. Accordingly, the performance of such schemes is appropriately described in terms of consecutive collision resolution intervals (CRIs).

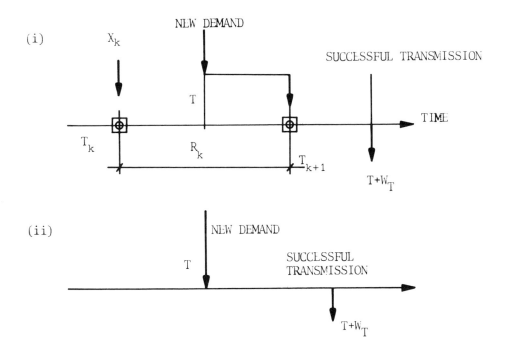

Fig.5. Temporary (i) blocked and (ii) free schemes

The outsets of the kth and (k+1)th CRIs, are denoted, respectively by T_k and T_{k+1}. X_k stands for the number of demands, getting access to the common medium from T_k on, and R_k for the length of the kth CRI. A demand arriving, say, at instant T, has to wait throughout the rest of the kth CRI until the instant T_{k+1}, to get an access to the common medium. However, the considered demand itself is met only at some later instant $T+W_T$, within the kth CRI, after eventually several transmission trials, according to the actually adopted collision resolution algorithm (CRA). As a matter of fact the packet delay, corresponding to the demand that arrived at T, is denoted by W_T.

In the most simple form of the temporary free case any new demand is immediately accessed to the common medium. However, the price of this promptness is an eventual increase in the number of already colliding demands, contending for successful transmission.

When considering this third kind of more sophisticated collision resolution schemes, it is usually reasonable to rely upon temporary slotted schemes and a slot length equal to the packet length; let us follow this practice in the sequel. For temporarily blocked as well as for temporarily free schemes some sort of feedback is needed near the end of each slot. Adopting either some simple binary feedback (collision, no collision) mentioned previously or the use of a ternary feedback (idle, collision, success) is of practical interest. By the end of each slot each of the transmitters just in collision may take one of either two or $Q > 2$ possible actions. Accordingly we call the CRA binary and Q-ary, respectively. (E.g., for a binary CRA the possible choices at each slot are: transmit or be silent). This choice has to be made either randomly, by flipping an unbiased (or biased) coin at the end of each slot, or in some other practically more appealing, equivalent way. E.g., for Poisson arrivals, usually assumed for serving an unlimited population of data users, registering and using the arrival time of the considered demand offers a practical alternative for this purpose.

Random access schemes are particularly suited for serving data users by more or less simple means. The most simple principle in this case is to transmit also the messages themselves during the CRI, and not just the identifiers of the two parties. Time is, of course, wasted in this case at any occasion when a successful transmission is not achieved immediately at the first trial. Obviously, less time is needed for the access if short packets convey identifiers, however the messages themselves are conveyed by long packets as all short packets already

reached the opposite party. More distinctly, the long
data packets are transmitted to their opposite parties
consecutively, according to either the instants of successful transmissions or the demand arrivals times.
(Information, in the former case, can be conveyed by the
short packet procedure.)

Random access as well as sequentially scheduled
schemes can be used also for serving jointly delay tolerant as well as delay sensitive users, e.g., data and speech
users, via the same common medium. This can be done by
embedding the stream of delay tolerant packets randomly
into a temporary periodically scheduled stream of active
delay sensitive packets.

The term active is meant, also in the present context,
in the usual sense. E.g., a 10 ms speech segment (and the
corresponding speech packet) is active during a talkspurt,
and otherwise silent.

COOPERATIVE QUEUING: AN INSIGHT INTO THE ANALYSIS

Next just some of the most basic aspects of analysing
the two service cooperative queuing task described in the
previous section, will be considered. We are going to
estimate (i) the maximum stable throughput and (ii) the
asymptotic average data packet delay (the latter, of
course, only for stable operation).

A temporarily slotted scheme, blocked by access and
a periodically scheduled speech transmission will be assumed, with slots and short packets of a unit length and
long packets of length K. The short packets are for data
access, the long packets for data and speech transmission.
(Fig. 6. Csibi 1986b, 1985, 1986a.)

Active speech packets of at most N simultaneously busy
talkers are admitted within a each frame of length M.
$(0 < NK < M)$ First the active speech segments, drawn
during the $(k-1)$th frame, are transmitted in the kth frame,
from the kth flag on afteranother. (Endpoints of slots are
called nodes, and those of the frames flags.) From the kth
flag on NK slots are optionally reserved for this purpose.
(Optionally means that the time reserved for the optional
transmission of the speech packet of the ℓth busy talker
is left free for data transmission, provided this talker
is just silent.) Short as well as long data packets are
transmitted within each frame during the time left free
by the actually transmitted (active) speech packets.
(More distinctly, first the short and next the long data
packets skip the slots already seized by the active speech

packets.

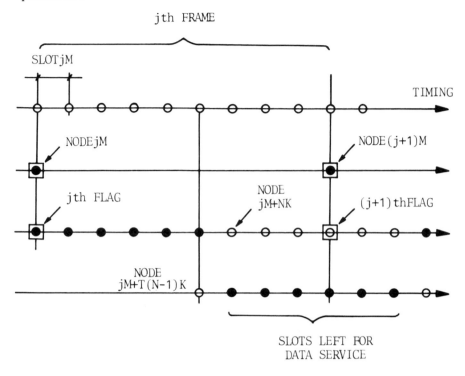

Fig.6. A way for joint data and speech transmission

Let us call kth block the time interval between the outset of the kth and (k+1)th CRI (including also T_k). The kth block includes the skipped active speech packets, the time needed for the transmission of X_k long data packets, and also some slots wasted for convenience (see Conditions(C.1) and (C.2) in the sequel). Denote by R_k the length of the kth block (Fig.5).

Introduce the following three constraints:

(C.1) skip slots next to the kth CRI of short data packets optionally, in order to start the transmission of long data packets always at a node, the distance of which from T_k is at an integer multiple of K.

(C.2) Skip also additional slots optionally, next to the transmission of long packets, in order to start the next block always at a flag.

(C.3) Let, during the kth block, active speech packets

be actually transmitted up to the Lth flag, next to T_k. Accordingly, let all active speech packets, occuring during the kth block next to the Lth flag, be lost.

Observe that the additional data packet delay due to Constraints (C.1) and (C.2) is obviously irrelevant w.r.t. the 250-1000 ms loop delay, still tolerable for data dialogs. The choice of the threshold L, according to (C.3), is irrelevant provided L is appropriately big. By the way, we will find a lower bound for the maximum stable throughput, in the sequel, that holds uniformly for any L, thus also for L → ∞. The next choice of the truncation threshold L will, therefore, finally not matter at least in the stability studies, for which the considered truncation is alone of interest.

Assume data demands to arrive according to a homogeneous Poisson process at a rate s_p, and that the on-off sequence $Y = (Y_k^{(\ell)}, k = o, 1, ...)$, describing the on-off state of the consecutive packets of the ℓth busy talker, is a stationary binary Markov chain with a positive transition probabilities $(0 < \ell < N-1)$. Assume that for $0 \leqslant \ell \leqslant N-1$, the on-off sequences $Y^{(\ell)}$ and also the data demand arrivals are independent. Assume the same activity for all ℓ; i.e.: $p = EY_o^{(o)} = EY_k^{(\ell)}$. Let $p > 0$. Call

$$S_D = s_D K, \qquad \bar{S}_S = \frac{NKEY_o^{(o)}}{M} \quad \text{and} \quad \bar{S} = S_D + \bar{S}_S$$

data throughput average speech throughput and average overall throughput respectively. ($EY_o^{(o)}$ is usually called the activity of the speech packets. E stands for the expectation.)

Adopt, for the sake of definiteness, random access for data demands with a branchwisely scanned version of the well known efficient symmetric tree algorithm, according to Sec. 1 in Tsybakov and Mikhalov (1978). Assume a propagation delay that is small w.r.t. unity. Accordingly, assume that bits conveying the feedback, additional bits for identifying the presence of active speech packets, etc., and also appropriate guard spaces can be inserted into any slot, next to each short packet: accordingly, include the time, necessary for these items, into the length of the short and long packets, respectively.

Obviously, in the present context, the sequence

$X = (X_k; \; k = 0,1,\ldots)$ of the number of data demands, accessing the common medium at $T_0, T_1, \ldots$, is no longer sufficient for describing the performance of the service, as, for an appropriate description, the impact of the number of skipped active speech packets during each block should also be taken into account.

It turns out however, that the two variate random variable (X_k, U_k) may already serve as an appropriate state variable. Here U_k is some appropriate descriptor of the on-off states of the NL speech packets, skipped at most during the kth block. Define, for convenience, U_k as a non-negative integer valued random variable, the binary representation of which is of finite bits, and the ℓth bit of this representation being $Y_k(\ell)$. As U_k is finite valued,

$$Z_k = cX_k + U_k \qquad (1)$$

can be considered instead of (X_k, U_k), $(c = \sum_{i=1}^{L} 2^{iN}.)$ Accordingly Z_k will be used, for the kth block, as a state variable, in the sequel.

It is relevant that, for the evolution of the state variable sequence $Z = (Z_k, \; k = 0,1,\ldots)$, the following constraint holds:

$$E(X_{k+1} \mid Z_k) < aX_k + V_k + b \qquad (2)$$

a and b stand for known non-negative integers, and V_k for a random perturbation, for which

$$r^{-1} \sum_{k=0}^{r} V_r \to 0 \qquad (3)$$

holds, as $r \to \infty$, almost surely.

By the Poissonian assumptions on the demand arrivals, the Markovian assumptions on the on-off sequences and also by Conditions (C.1) and (C.2), the state variable sequence Z turns out to be homogeneous, irreducible, (aperiodic) Markovian. It is also relevant that the state variable Z_k is of the form (1), where X_k is countable, however V_k is only finite valued.

For a Markovian state variable sequence Z it is sufficient to identify the stability of the performance simply

with the ergodicity of Z. However as the conditioning in
(2) is w.r.t. Z_k (and not w.r.t. to X_k) and as (2) also
contains the perturbation term V_k, Foster's classic ergodicity theorems for Markov chains, so appropriate for the
study of blocked access for pure data transmission, no
longer suffice. However by extending these theorems to a
controlled Markov sequence Z with a finite valued control
sequence U, and with long run perturbations V_k in (2)
(decaying according to (3), as $k \to \infty$) one can derive, also
in this case, a simple and useful ergodicity condition.

It turns out, in this way, that Z is stable for any
a < 1. This happens if the overall average throughput $\bar{S}$
is   less   than $\bar{S}_M$, the supremum of the stable average
overall throughput. A simple lowerbound $\bar{S}_\infty$ on $\bar{S}_M$ is obtained (uniformly for any L, thus also for $L \to \infty$) of the
following form:

$$\bar{S}_\infty = \frac{1+\bar{S}_S(h/K)}{1+(h/K)} \tag{4}$$

(h = 8/3, Fig. 7).

For a long packet length K = 30 and an average speech
throughput $\bar{S}_S$ = 0.6. ($\bar{S}_\infty(\bar{S}_S) < \bar{S}_L < 1.$)

For throughput pairs $\bar{S}_S$ and S_D, for which the state
variable sequence Z is ergodic, the asymptotic average
delay

$$D = \lim_{t\to\infty} E(W_T|T=t) \tag{5}$$

exists. More distinctly, D is overestimated, following
essentially the same approach, and with essentially the
same tightness, as by Tsybakov and Mikhalov (1978) for
pure data transmission (Csibi 1986 b and 1985).

We confined ourselves, in the present section, just
to giving a brief insight into the stability and delay
analysis of a typical joint service task. We investigated neither buffering of the consecutive demands of a
user nor the input and output policies to be related to
this (Ephremides and Zhu 1986, Lim and Meerkov 1986,
Mathys and Falting 1986). The impact of channel errors
(Tsybakov 1985, Vvedenskaya and Tsybakov 1983a, and
Massey 1981) and that of capturing in frequency modulated
terrestrial radio systems (Sidi and Cidon 1985) were not

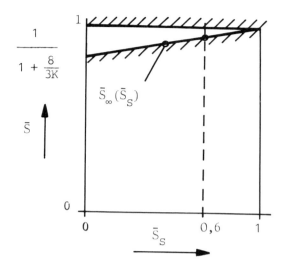

Fig.7. A lower bound $\bar{S}_\infty(\bar{S}_S)$ for the supremum $\bar{S}_M$ of the stable overall average throughput $\bar{S}$ versus the average throughput $\bar{S}_S$ ($\bar{S}_\infty$ ($\bar{S}_S$) < $\bar{S}_M$ < 1.)

treated either. The task of serving speech and data jointly was considered without any environmental constraint. However, adding a speech service to an existing random access data service without any changes at the place of the already operating data work stations, is also a topic of much interest. So doing is of particular significance for existing local area networks with carrier sensing, widespread examples of which are the ETHERNET networks (Fratta and Szabo 1986).

MULTI-USER CODING: SOME INSIGHT INTO PRINCIPLES AND ANALYSIS

There are two basic kinds of coding tasks of particular recent interest in multiple-access communications. One of these offers further possibilities mainly for terrestrial radio and satellite relay systems by code division: the trade off is the excessive use of the frequency spectrum. The other has got promising perspectives for random access techniques without feedback: the price is an extended use of time.

The common medium is no longer confined, in these cases, to serving at most one user at a time. As a matter of fact the essential common feature of both approaches is that users are enabled to convey messages to their opposite parties through the common medium, with a fair reliability, at any time. The codes are devised according to this purpose: the terms code division, signature coding, superimposed codes, used in this context, just emphasize the different features such codes exhibit.

As is well known, two basic kinds of spread spectrum schemes exist: (i) the direct sequence schemes, using appropriately distinguishable (viz., in a sense, almost orthogonal) symbol sequences for coding each bit of the source sequence individually, and (ii) frequency hopping schemes with some appropriate two dimensional time and frequency coding of either the individual source bits or a block of source packets. One has got for frequency hopping still a further subdivision: slow frequency hopping with a hopping rate much below and fast frequency hopping with a hopping rate much above the source bit rate.

At least three features make spread spectrum schemes more and more appealing also in Public Telecommunications and related fields. Viz. :

— the relative insensitivity to multipath propagation and radio interference, a feature of particular interest for terrestrial radio systems

— the remarkable freedom users obtain in this case to access each other, cellular nodes or satellite relays; a feature mainly due to the code-division used, and to the flexibility of distributed packet switching

— last but not least, the very fact that continuously performing as well as bursty users can be equally well accessed to the common medium in this case; the access scheme no longer relies so intimately on the specific character of the users.

The price for these features is of course the much broader frequency band needed on a route as in FDMA. This does not mean, however, that the spectrum occupied by a large network is still so excessive, as much can be saved, in this case, by the interference robustness and the flexible network design.

A tour de force not only for terrestrial radio systems but also for geostationary satellite relay systems is to withstand the interference due to near transmitters when the opposite party is far away. This is the so called near-far effect. Frequency hopping lends itself to this purpose. An excessive use of the frequency spectrum can more probably be avoided by using slow and not fast frequency hopping in these cases (Pursley 1981 and 1986, Geraniotis and Pursley 1982, Pickholz et al. 1982).

Errors in frequency hopping are mainly due to hitting those time-frequency cells which are just occupied by the ones of the user of interest. However the temporary collision of contending packets usually causes such a hit only with a probability much less than unity. The conditional probability of a packet error, given a collision, can be further diminished by using some appropriate error correction code. Delay tolerant data packets need not be lost even in this case: they can be saved by adopting some sort of random access scheme for this purpose. As packet errors usually occur very infrequently, even simple random access schemes, e.g. some variant of the ALOHA scheme, can successfully be employed in this case.

Obviously, a number of significant and interesting channel modelling and coding questions can be posed in this context. It is, therefore, not surprising that so much activity has been invested in this respect recently (Einarsson 1980, Ericson 1985, Vajda 1983, Györfi and Vajda 1983, Györfi and Kerekes 1983 and Molnar and Vajda 1984).

For the other kind of multi-user coding, meant for random access without feedback, the basic idea is to send the source bits only at selected positions of the coded sequence, giving a chance to the contending sequences not to hit each other. Much time is, of course wasted in this case. One usually picks blocks of sequences of length N for this purpose. The first question is: How many simultaneously busy users can be served, in the case of appropriate coding, simultaneously, within a delay N? A second, more practical question is what delay N is needed if just any subset of M active users of a user population of size T is to be served simultaneously. The problem is that N increases too rapidly with T (for given N/T) unless some numerical care is taken. Anyhow, for given N, (T,M) pairs are known for solving the problem for certain, relatively simple cases. What is surprising is that the throughput of the slotted ALOHA scheme can be achieved, also under such stringent conditions, even without any slotting (Massey and Mathys 1985, Tsybakov and Likhanov 1983, Massey 1982, Huber and Shah 1975, Nguyen et al. 1985, Dyachkov and Rykov 1983, Nguyen and Zeisel 1985, Nguyen 1986).

REFERENCES

Abramson, N. (1970). The Aloha System - another alternative for computer communication. in APIPS Conf. Proc., 37, FJCC, 695-705.

Abramson, N. (1985). Development of the ALOHANET. IEEE Trans. IT-31, 119-123.

Bassalygo, L.A. and Pinsker, M.S. (1983). Limited multiple access of a nonsynchronous channel. Probl. Peredachi Inform., 19, 92-96.

Borogonov, F., Fratta, L., Tarini, F., Zini, F. (1985). L-EXPRESSNET: The communication subnetwork for the C-net project. IEEE Trans. COM-33, 612-619.

Capetanakis, J.I. (1979a). Tree algorithms for packet broadcast channels. IEEE Trans. IT-25, 505-515.

Capetanakis, J.I. (1979b). Generalized TDMA: The multi-accessing tree protocol. IEEE Trans. COM-27, 1476-1484.

Csibi, S. (1985). On the stability of random access data communication during the time left by speech packets. Probl. Contr. Inform. Theory. 14, 231-246.

Csibi, S. (1986a). Extending Foster's ergodicity criteria to controlled Markov chains and analyzing integrated service local area networks. In Trans Tenth Prague Symp. Inform. Theory, Statist., Dec. Funct., Rand. Proc. (in press).

Csibi, S. (1986b). More on the stability of a data packet flow embedded ramdomly into a partially active speech packet flow. Preprint, TU of Budapest. (Probl. Contr. Inform. Theory, 15, 4, 1987. in press.)

Dallos, Gy. (1978). A modified carrier sence multiple-access procedure for interactive radio terminals. In Proc. Sixth Coll Microw. Com., OMDK Budapest, I-3, 17.1 - 17.3.

Dyachkov, A.G. and Rykov, V.V. (1983). Survey of superimposed code theory. Probl. Contr. Info. Theory. 12, 229-244.

Einarsson, G. (1980). Address Assignment for time-frequency coded spread-spectrum system. BSTJ. 59, 1241-1255.

Ephremides, A. and Zhu R-Z (1986). Delay analysis of interacting queues with an approximate model. Preprint, EE,U. of Maryland. 1-22.

Ericson, T. (1985). Frequency hopping multiple-access system and protocol sequence. Tech. Rept. Lingkoping U.

Fratta, L. and Szabo, Cs. (1986). CSMA/CD compatible protocols to support real-time voice transmission. Tech. Rept. LCE-86-1, Dep. de Electronica, Polit. di Milano.

Gallager, R.G. (1985). A perspective on multiaccess channels. IEEE Trans. IT-31, 124-142.

Georgiadis, L. and Papantini-Kazakos, P. (1985). Limited feedback sensing algorithms for the packet broadcast channel. IEEE Trans. IT-31, 280-294.

Geraniotis, E.A. and Pursley, M.B. (1982). Error probabilities for slow-frequency-hopped spread-spectrum multiple-access communications over fading channels. IEEE Trans. COM-30, 996-1009.

Green, P.E., Jr. (1984). Computer communications: milestones and prophecies. IEEE Com. Mag., 22, 49.63.

Green, P.E., Jr. and Godard, D.N. (1986). Prospects and design choices for integrated private networks. Preprint, R_m H3D8 Research Yorktown, 1-18. (IBM Syst. J., invited paper, in press.)

Györfi, L. and Kerekes, J. (1981). A block code for noiseless asynchronous multiple access OR channel. IEEE Trans. IT-27, 788-791.

Györfi, L. and Kerekes, I. (1983). Analysis of multiple-access channel using multiple level FSK. In Cryprography (Beth, T., ed.). 165-197. Springer-Verlag.

Györfi, L. and Vajda, I. (1983). Block coding and correlation decoding for an m-user weighted adder channel. Probl. Contr. theory. 12, 405-417.

Huber, J. and Shah, A. (1975). Simple asynchronous multiplex system for unidirectional low-data-rate transmission. IEEE Trans. COM-23, 675-679.

Hung, J.C. and Berger, T. (1985). Delay analysis of interval searching contention resolution algorithms. IEEE Trans. IT-31, 264-274.

Kleinrock, L. (1985). On queueing problems in random access communications. IEEE Trans. IT-31, 166-175.

Kleinrock, L. and Tobagi, F.A. (1975). Packet switching in radio channels. Pt. I. IEEE Trans. COM-23, 1400-1416.

Lazar, A.A., Patir, A., Takahashi, T., and El Zakri, M. (1985). MAGNET: Columbia's integrated network testbed. IEEE J. Select. Areas Com. SAC-3, 859-871.

Lazar, A.A. and White, J.S. (1986). Packetized video on MAGNET. Preprint. Center f. Telecom. Res., Columbia U. 1-10.

Lim. J.T. and Meerkov, S.M. (1986). Theory of Markovian access to collision channels. Pts. I-III. Techn. Rept. Nos. 235-237. Com. Sign Proc., U. Mich., 1-27.

Massey, J.L. (1981). Collision resolution algorithms and random access communications. In Multi-User Communications (Longo, G., ed.) CISM Courses and Lectures, No. 265, 73-137. Springer--Verlag, Wien-New York.

Massey, J.L. (1982). The capacity of the collision channel without feedback. In Abstract of Papers. IEEE Int. Symp. Inform. Theory. Les Arcs, France, June 21-25.101.

Massey, J.L. and Mathys, P. (1985). The collision channel without feedback. IEEE Trans. IT-31, 192-204.

Mathys, P. and Falting, B.V. (1986). The effect of channel exit protocols on the performance of finite population random access systems. Preprint. Dept. ECE, U. of Colorado.

Mathys, P. and Flayolet, P. (1985). Q-ary collision resolution algorithms in random access systems with blocked channel access. IEEE Trans. IT-31, 217-243.

Molnar, L. and Vajda, I. (1984). Decoding error probability of the Einarsson-code for a frequency-hopped multiple-access channel. Probl. Contr. Inform. Theory. 13, 109-120.

Nguyen, Q.A. (1986). Some coding problems of multiple-access communication systems. Tech. Rept., HEI, TU of Budapest.

Nguyen, Q.A., Györfi, L. and Massey, J.L. (1985). Some constructions of protocol sequences for a collision channel without feedback and a class of cyclic constant weight codes. Preprint. Inst. f. Sign. Proc., ETH Zurich and HEI, TU of Budapest.

Nguyen, Q.A. and Zeisel, T. (1985). Packet communication on a T user collision channel without feedback: cyclotomic protocol sequences. Trans. COMNET '85. 8/40-53.

Parr, F.N. and Green, P.E. (1985). Communication for personal computers. IEEE Com. Mag. 23, 26-36.

Pickholz, R.L., Schilling, D.L. and Milstein, L.B. (1982). Theory of spread spectrum communications. A tutorial. IEEE Trans. COM-30, 855-885.

Pursley, M.B. (1981). Spread spectrum multiple-access communications. in Multi-user Communication Systems. (Longo, G., ed.) CISM Courses and Lectures. No. 265. 139-199. Springer-Verlag, Wien-New York.

Pursley, M.B. (1986). Frequency-hop transmission for satellite packet switching and terrestrial packet radio networks. IEEE Trans. IT-31, 143-165.

Rao, R. and Ephremides. A. (1986). On the stability of interacting queues in a multiple-access system. Preprint, EE, U. Maryland.

Syka, E.D., Karvelas, D.E. and Protonotarios, E.N. (1986). Queueing analysis of some buffered random multiple-access schemes, IEEE Trans. COM-34, 790-301.

Sidi, M. and Cidon, I. (1985). Splitting protocols in presence of capture. IEEE Trans. IT-31, 295-301.

Sidi, M. and Segall, A. (1983). Two interferring queues in packet radio networks. IEEE Trans. COM-31, 123-129.

Saadavi, T. and Ephremides, A. (1984). Analysis, stability and optimization of slotted ALOHA with a finite number of buffered users. IEEE Trans. AC-26, 680-689.

Tsybakov, B.S. (1985). Survey of USSR contribution to random multiple-access communications. IEEE Trans. IT-31, 143-165.

Tsybakov, B.S. and Fedortsov, S.P. (1986). Packet transmission by an unmodified blocked stack algorithm in random multiple-access communications. Probl. Peredachi Inform. 23, 96-102.

Tsybakov, B.S. and Likhanov, N.B. (1983). Packet communication on a channel without feedback. Probl. Peredachi Inform. 19, 69-84.

Vvedenskaya, N.D. and Tsybakov, B.S. (1983). Random multiple access of packets in a channel with errors. Probl. Peredachi Inform., 19, 52-68.

Vvedenskaya, N.D. and Tsybakov, B.S. (1984). Packet delay for multiple-access stack algorithm. Probl. Peredachi Inform. 20, 89-101.

Vajda, I. (1983). A coding rule for frequency hopped multiple-access channels. Probl. Contr. Inform. Theory. 13, 331-335.

Wolf, J.K. (1985). Born again group testing: multiaccess communications. IEEE Trans. IT-31, 185-191.

4

Coherent optical fiber communications

T. OKOSHI

ABSTRACT

Research and development of coherent optical fiber communications started late in the 1970s. Since then these have been expanded rapidly because of the expectation for repeater-separation elongation reaching 100km and superwideband frequency-division multiplexing (FDM) with very fine frequency separation (typically 10-100GHz). In this paper, the technical significance, history of research, and classification of various coherent optical fiber communication schemes are described first. In the latter half of the paper, recent progress in this area is reviewed, and the ultimate performance to be realized in the future is speculated.

1. INTRODUCTION

The present optical fiber communications are in a sense as primitive as the radio communications prior to 1930. The reason is that both of these abandon the phase information of the carrier; in other words, both of these are noncoherent (noise-carrier) communications.

The standard modulation/demodulation scheme being employed in the present optical fiber communications is called the intensity-modulation/direct-detection (hereafter IM/DD) scheme. The term "DD" stems from that the signal is detected directly at the optical stage of the receiver; neither frequency conversion (heterodyne or homodyne scheme) nor sophisticated signal processing at lower frequencies is performed.

On the other hand, in the history of radio communications, the heterodyne scheme has become common since 1930, and is now widely used even in pocket radio receivers. Sophisticated coherent modulations such as FM, PM, frequency-shift keying(FSK), and phase-shift keying(PSK) are also widely used in broadcasting and communications.

Here a question arose late in the 1970s: will the IM/DD system continue to be predominant in optical communications?

Or will it gradually retire to yield its position to more sophisticated heterodyne/homodyne and/or coherent systems?

The IM/DD system has a great advantage in system simplicity and low cost. On the other hand, some applications of optical fiber communications exist in which a long repeater separation is our primary concern; an example is the undersea optical fiber communications between islands. In such a case, improvement of the equivalent receiver sensitivity by a heterodyne-type receiving technique, or by a coherent modulation/demodulation scheme such as PCM-FSK or PCM-PSK may become advantageous, even at the sacrifice of simplicity and low cost.

Especially noteworthy is that the sensitivity improvement is particularly dramatic (10-25dB) at the 1.5-1.6μm wavelength region, where the silica-fiber loss becomes minimum ($\lesssim$ 0.2dB/km) whereas good photodetectors (for the DD scheme) are not available. The sensitivity improvement of 20dB will result in a repeater-separation elongation of 100km when the fiber loss is 0.2dB/km.

The above expectation on receiver-sensitivity improvement is the principal motivation behind the present worldwide efforts toward heterodyne/coherent optical fiber communications. The second expectation is the high frequency-selectivity offered by the heterodyne scheme. This will be useful in superwideband telecommunications as well as in superwideband local area networks (LANS) or subscriber networks or high definition CATV networks, but the latter three will be realized probably in a relatively distant future.

2. HISTORY OF RESEARCH

Papers on the heterodyne/coherent optical fiber communications began to appear in 1979 in Japanese (Okoshi, 1979) (Yamamoto, 1979) and in 1980 in English (Okoshi and Kikuchi, 1980) (Favre and LeGuen, 1980) (Yamamoto, 1980).

Since then the research in this area has been expanded rapidly year by year; most of the major communications laboratories in the world have started the research and development in the past eight years; ECL (Japan) in 1979, CNET (France) in 1980, BTRL (UK) in 1981, NEC Res. Labs. (Japan) in 1982, HHI (Heinrich Hertz Inst.; DBR) in 1982, Bellcore (USA) in 1984, and AT&T Bell Labs. in 1984-1985. Coherent optical fiber communications is now one of the major topics at international conferences on optical communications.

3. DEFINITION AND CLASSIFICATION OF COHERENT OPTICAL COMMUNICATION SCHEMES

Radio-frequency or optical receivers are classified into three basic categories: direct-detection receivers, heterodyne receivers and homodyne receivers.

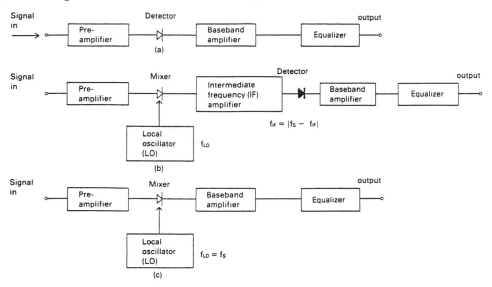

Fig.1 Three basic constructions of radio-frequency or optical receivers: (a) direct-detection receiver, (b) heterodyne receiver, and (c) homodyne receiver.

In a direct-detection receiver (FIG. 1(a)), the received radio-frequency or optical signal (modulated carrier) is directly converted to the baseband signal. Sometimes (in most radio receivers but very seldom in optical receivers) the signal is amplified before it is detected. In a heterodyne receiver (Fig.1(b)), the received signal is mixed with the so-called local oscillator (LO) power. The difference-frequency signal obtained, called intermediate frequency (IF) signal, is then amplified and detected. Note that $f_{IF} = |f_s - f_{LO}|$. In a homodyne receiver (Fig. 1(c)), the LO frequency as well as its phase are controlled so that they are always equal to the frequency and phase of the received signal carrier. A critical technique such as the phase-locked loop (PLL) scheme must be employed to achieve the matching of the frequency and phase.

In the following, the term "heterodyne" is sometimes used for simplicity in a wide sense including the "homodyne," because the latter is a special case of the former.   However, in many cases (for example in the discussion of the bit-error rate) distinction between heterodyne and homodyne is essential.

The term "coherent" in the title of this paper also needs comments.   Presently the term "coherent" is used in two different meanings.   In the first meaning, those cases in which the carrier is intensity-modulated (hereafter IM) as a noise carrier are not included in the "coherent" schemes,   because the temporal coherence of the carrier is not utilized.   Hence,   for example,   a PCM-OOK (on-off keying) heterodyne system is classified in "noncoherent heterodyne" systems, whereas a PCM-PSK (phase-shift keying) heterodyne system is classified in "coherent heterodyne" systems.   In the second meaning,   the PCM-OOK heterodyne system is also included in the "coherent" class, because on the surface of the frequency-mixing diode in this system, we take advantage of the spatial coherence of the carrier. Thus in the second meaning,   all heterodyne and homodyne schemes are coherent schemes.

We may classify all optical communication systems into four classes as shown in Table 1.   In the title of this paper, the term "coherent" is used in the second meaning; i.e., we consider that all heterodyne and homodyne schemes (i.e., the lower half of Table 1) are the subjects of this paper.

TABLE 1 Classification of various optical fiber communication schemes. The term "coherent" is used in
its narrow sense in this table.

Receiver front end	Modulation/Demodulation	
	Non-coherent	Coherent
Non-heterodyne (Direct detection)	PCM-IM/DD (Intensity-Modulation/Direct-Detection)	Realizable, but no technical advantage
Heterodyne (including homodyne)	PCM-OOK (PCM-IM) PCM-FSK (with non-coherent detection)	PCM-ASK PCM-FSK PCM-PSK

4.   IMPROVEMENT OF RECEIVER SENSITIVITY BY COHERENT SCHEMES

4.1 Two Sensitivity-Improving Effects

The greatest advantage of a coherent system is the improvement of the receiver sensitivity, that is, the reduction of the minimum receiving signal level for achieving a prescribed bit-error rate (hereafter BER), for example, BER=10^{-9}.

This improvement is attributed to two effects (Okoshi et al., 1981), as shown in Fig.2 (Okoshi, 1984). One is the improvement of the S/N at the output end of the receiver preamplifier (for a given signal power P_s) by the use of the heterodyne or homodyne scheme (the left half of Fig.2). The other is the improvement of the BER (for a given S/N) brought about by the use of a coherent modulation/demodulation scheme as compared with a noncoherent scheme (the right half of Fig.2).

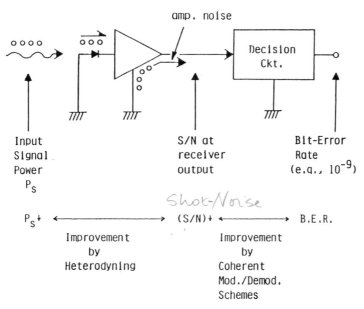

amp. noise

Decision Ckt.

| Input Signal Power P_s | S/N at receiver output | Bit-Error Rate (e.g., 10^{-9}) |

Shot-Noise

$P_s\downarrow$ ⟷ $(S/N)\downarrow$ ⟷ B.E.R.

Improvement by Heterodyning

Improvement by Coherent Mod./Demod. Schemes

Fig.2 Receiver-sensitivity improvement by heterodyne/ coherent schemes (Okoshi, 1984).

4.2 Shot-Noise Limit

In an optical detection, i.e., in an opto-electronic (OE) signal conversion, an absolute S/N-limitation exists due to the fact that light is not continuous but consists of photons, which produce a flow of discrete electrons at the electrical terminals of the photodetector. As the result, an ideal photodetection is not a noise-free detection but a shot-noise-limited detection, giving

$$(S/N)_{ideal} = P_s/2hfB, \qquad (1)$$

where P_s denotes the received signal power, h the Planck constant (=6.625×10^{-34} J.s), f the optical signal frequency, and B the bandwidth of the receiver.

In a direct-detection receiver, due to the dark-current noise of APD and the noise from the following baseband amplifier, the S/N is usually very much deteriorated from Eq.(1). An S/N close to Eq.(1) (with a few dB difference) can be obtained only at wavelengths below 1 μm where a Si-APD can be used. The S/N is much deteriorated at longer wavelengths where a Ge APD or III-V compound APD must be used.

4.3 Sensitivity Improvement by Heterodyning

The heterodyne/homodyne scheme is presently the only practical method with which a nearly shot-noise limited detection can be realized at 1.3μm or 1.55μm.

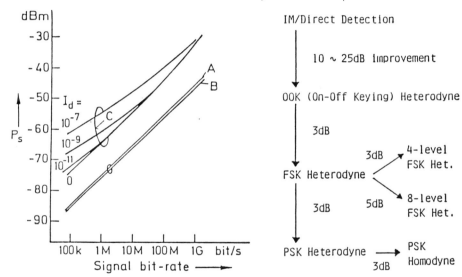

Fig.3 The received signal level for achieving BER=10^{-9} as functions of signal bit-rate (Okoshi et al., 1981).

Fig.4 Receiver-sensitivity improvement by various heterodyne (or homo-dyne) and coherent modulation/demodulation schemes.

Figure 3 shows an example from the results of analysis comparing the IM/DD and ASK/heterodyne systems (Okoshi et al., 1981). The ordinate shows the received signal level necessary to obtain a BER of 10^{-9}. The abscissa is the bit rate of the PCM signal. The lower two curves show the required signal levels for ASK/heterodyne cases: curve A for envelope (noncoherent) detection and B for coherent detection. The upper four curves show the required signal levels for IM/DD cases. The excess noise factor x of the APD is assumed to be 1.0, corresponding to long-wavelength (1.3-1.6μm) Ge detectors. The parameter I_d denotes the dark current in amperes.

Figure 3 indicates that the improvement in the required signal level, from the IM/DD scheme to the ASK/heterodyne scheme, ranges between 10 and 25dB. Generally speaking, the advantage of heterodyne detection is emphasized at long wavelengths (1.3-1.6μm) where x and I_d are large.

4.4 Sensitivity Improvement by Coherent Modulation/ Demodulation

The technical significance of the heterodyne scheme lies partly in that it is a premise for using coherent modulations such as FSK or PSK. The sensitivity improvement by these schemes had been known in general communications theory (Stein and Jones, 1965) when it was modified and applied to coherent optical communication systems (Yamamoto, 1980) (Okoshi et al., 1981).

The sensitivity improvement by various schemes are summarized in Fig.4. By the use of ASK/heterodyne scheme, 10-25dB improvement can be expected as compared with the IM/DD scheme, as has been found in Fig.3. As compared with ASK/heterodyne, 3-dB further improvement can be expected with FSK/heterodyne scheme, and again 3-dB further improvement with a PSK/heterodyne scheme. Still further 3-dB improvement can be obtained with a PSK/homodyne system.

On the other hand, 3-dB and 5-dB improvement could be expected if we realize 4-level and 8-level FSK systems, respectively. This is possible bacause, in such cases, we can reduce the signal bit-rate by 1/2 and 1/3, respectively, still transmitting the same amount of information. Note that the reduction of the signal bit-rate brings forth an improvement in the sensitivity because the receiver bandwidth can be narrowed.

5. FREQUENCY-SELECTIVITY IMPROVEMENT BY HETERODYNING

The second advantage of a heterodyne system is that the optical frequency selectivity can be improved because of the good frequency selectivity of the intermediate frequency(IF) amplifier. Thus, frequency-division multiplexing (FDM) with very fine carrier separation becomes possible.

However, in the present state-of-the-art, this advantage cannot be emphasized for a long-haul telecommunication system, because the optical power loss in the optical multiplexer at the transmitting end and demultiplexer at

the receiving end will not be negligible, when the
frequency separation between carriers is small.

Figure 5 illustrates the problem. Figure 5(a) shows an
example of the carrier-frequency allotment in a superwideband FDM system described in one of the earliest papers
(Okoshi, 1979). In this case ten carrier frequencies are
arranged with 100GHz separation in each carrier group
(corresponding to the passband of an optical branching
filter) having 2THz bandwidth. Figure 5(b) shows how the
ten carriers are divided into ten heterodyne receivers. If

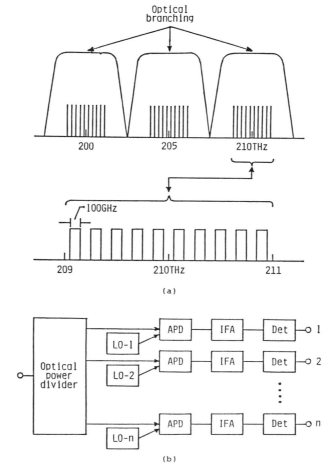

Fig.5 (a) An example of carrier frequency allotment
         in an FDM system using heterodyne receiving
         technique.
      (b) Power-dividing circuit in a heterodyne FDM
         receiver (Okoshi, 1979).

the power divider in Fig.5(b) has no frequency selectivity, all the signals will undergo an appreciable branching loss, i.e., at least 10-dB loss when ten carriers are divided into ten receivers.

Such a branching loss is more or less inevitable in the present state-of-the-art. However, research is now in progress; in the near future, optical branching filters with 100GHz separation (or even 10GHz separation) will become practical (Okoshi, 1986).

On the other hand, we may take advantage of the fine frequency selectivity of the heterodyne system in some applications even within the present state-of-the-art; these applications are superwideband local area networks (LANs), CATV networks, and integrated-services digital networks (ISDN) to be realized in the future. Note that the multiplexing or demultiplexing loss is less serious in such short-haul systems.

6. RELEVANT TECHNICAL TASKS

The following technical problems must be overcome before a practical coherent systems are realized:

(1) The most important problem is how to stabilize the frequency of semiconductor lasers, which will most possibly be used as the transmitter as well as the local oscillator (LO). Since the typical IF frequency is 0.2 -2GHz, which is about 10^{-6}-10^{-5} times the signal frequency (typically 200THz), the requirement for the frequency stability is very severe.

(2) The spectral purity of the semiconductor laser must also be improved. If the carrier or the LO output is noisy, the frequency or phase fluctuations will deteriorate the bit-error rate, thus increasing the required signal level.

(3) Practical, simple, and stable ASK, FSK, and PSK modulation/demodulation techniques must be developed.

(4) Practical, stable, low-noise heterodyne or homodyne receivers must be developed.

(5) Random fluctuation of the mixer efficiency due to fluctuation of the polarization state of the received optical signal must be prevented. This becomes possible by the use of a polarization-maintaining fiber, or by some other appropriate means such as a polarization controlling device at the input end of the receiver and/or a polarization diversity scheme.

(6) Although its technical merit has not yet fully been
proved, a laser preamplifier should be investigated
because it might improve the overall performance of
coherent communications in the future

(7) Finally, combining some of or all of the above techniques, experimental systems must be constructed to
demonstrate the aforementioned advantages in a practical scale, and to prove the technical feasibility of
coherent optical communications.

7. **REVIEW** OF SYSTEM EXPERIMENTS PERFORMED IN 1981-1985

Table 2 tabulates 30 system experiments reported in the
past five years, describing bit-error rate (BER) measurements of PCM-ASK, PCM-FSK, PCM-PSK, and PCM-DPSK systems.
Columns designated as 1985 (S:Spring) and 1985 (F:Fall)
show those papers delivered at OFC'85 in San Diego and
100C/ECOC'85 in Venice, respectively. The technical details of the papers in and before mid-1983, from late 1983
through 1985(S), and in 1985(F), are described in (Okoshi,
1984), (Okoshi, 1987), and (Okoshi, 1986), respectively. A
remarkable trend found in this table is that the
researchers' concern seems to have moved from ASK to the
more sophisticated FSK, PSK, and DPSK.

TABLE 2 Bit-error rate measurements of various coherent
communication system reported in 1981-1985.
Columns designated as 1985 (S:spring) and 1985
(F:fall) show those papers mainly read at OFC'85
and 100C-ECOC'85, respectively.

	1981	1982	1983	1984	1985 (S)	1985 (F)
ASK	U.Tokyo	BTRL U.Tokyo BTRL	BTRL	NEC		
FSK		ECL(M)	ECL(M)	HHI NEC BTRL	BCR ECL(Y)	NTT NEC BCR
PSK			BTRL U.Tokyo BTRL BTRL	HHI TU Wien		
DPSK		CNET	NEC BTRL BTRL		NEC NEC	AT&T-B NEC

The longest repeater separation reported so far is
301 km reported by researchers of NEC Corporation in 1986.
In early 1987, the achievements in such long-distance

systems were summarized and reported (Minemura, 1987). Some of his graphs are reproduced in a textbook (Okoshi and Kikuchi, 1987).

8. ULTIMATE RECEIVER SENSITIVITY AND REPEATER SEPARATION

The ultimate repeater separation of an optical communication system is determined by (1) maximum transmitter power, (2) fiber loss, and (3) receiver sensitivity. In the following discussion, we consider light transmission at 1.55μm, and assume that the fiber has a transmission loss of 0.2dB/km including splicing.

(1) Maximum Transmitter Power

The maximum transmitter power is limited by various nonlinear phenomena in optical fibers. The fiber nonlinearity is becoming important because the maximum power delivered by semiconductor lasers is now approaching 100mW.

The nonlinear phenomena in silica fibers are (1) stimulated Raman scattering (SRS), (2) stimulated Brillouin scattering (SBS), (3) self phase modulation (SFM), and (4) four-wave parametric mixing (Stolen, 1980). In coherent optical fiber communications which use narrow linewidth lasers and a single-mode fiber, the SBS effect is most critical.

It is known that the threshold power for the SBS effect is proportional to the linewidth of laser light, and depends critically upon the spectral shape due to modulation. For a purely single-frequency signal, nonlinearity due to the SBS effect is observed for input powers greater than several milliwatts, for both transmission and reflection. However, when the laser oscillates at a multifrequency mode, nonlinearity may not be observed up to 100mW (Cotter, 1983).

It is predicted that the SBS effect can be suppressed greatly (typically by 20dB) by proper choices of laser spectral shape and modulation scheme(cotter, 1983). In the following, we tentatively assume that a launched power of 10mW is mostly tolerated, and that of 100mW will become possible in the near future.

(2) Required Number of Photons per Signal Bit

Table 3 shows the required number of photons per one signal bit for various coherent optical communication schemes to achieve bit-error rate (BER) of 10^{-9} (Okoshi, 1986). It is found that the required number ranges between

10 and 80 for various coherent schemes.

It is interesting to compare these numbers with actually achieved data. Table 4 summarizes some of the data from BER measurements reported at IOOC/ECOC'85 at Venice, Italy, in October 1985. It is found that the actual photon numbers are several times greater than the ideal values. However, note that N=1000 or even greater in ordinary well-designed intensity-modulation direct-detection (IM-DD) receivers.

TABLE 3
Number of photons per one signal bit required to achieve BER=10^{-9}

Coherent schemes		$N_{required}$
Heterodyne	ASK	80
	FSK	40
	PSK/DPSK	20
Homodyne	ASK	40
	PSK	10

TABLE 5
Ultimate repeater separation in coherent optical fiber communications.

Signal bit-rate	Transmitted power		
	1mW	10mW	100mW
1Mb/s	450km	500km	550km
10Mb/s	400km	450km	500km
100Mb/s	350km	400km	450km
1Gb/s	300km	350km	400km

TABLE 4 Bit-error rate measurements reported at IOOC-ECOC'85, Venice, Italy

Organization	NTT	AT&T Bell Labs.		NEC
System	FSK Heterodyne	DPSK Heterodyne	DPSK Heterodyne	FSK Heterodyne
Wavelength	1530nm	1500nm	1500nm	1550nm
Bit rate	400Mb/s	400Mb/s	1Gb/s	140Mb/s
Distance	251km	150km	150km	243km
Devices TX	LD+Mirror	LD+Mirror	LD+Mirror	Two-electrode LD
Mod	Direct	$LiNbO_3$	$LiNbO_3$	Direct
Fiber loss	0.2dB/km	0.3dB/km	0.3dB/km	0.22dB/km
type	Normal	Disp.shift	Disp.shift	Normal
Photons/bit	240	90	270	350

(3) Ultimate Repeater Separation

At the wavelength of 1.55μm, a launched power of 100mW corresponds to 3.76×10^{18} photons per second. Hence, assuming the fiber loss as 0.2dB/km and the required photon number as 37.6 to facilitate computation (approximately the required number for FSK-heterodyne or ASK-homodyne cases), we obtain the maximum repeater separations for various combinations of signal bit-rate and transmitter power as shown in Table 5. These fantastic distances might look ridiculous at present, but will become realistic sometime in the future when technological difficulties are overcome.

Frequently asked questions: When and/or where will the coherent optical fiber communications featuring such a long-distance transmission capability be first put to practical use? It is difficult to answer the question "when." However, as to the question "where," one of the natural answers is that it will be realized first in inter-island (or island-continent) undersea communications. The comparison of the conditions in the Atlantic and Pacific Oceans is interesting in this respect. Investigation of the system lengths of existing undersea communications channels in the two oceans reveals that there are a lot of 100-400km systems in the Atlantic, whereas in the Pacific most of the systems are longer than 1000km (Mochizuki, 1986). This means that the Atlantic Ocean has more places than the Pacific Ocean does where coherent communications will make the use of undersea repeaters unnecessary.

9. SUMMARY

Recent progress in research on coherent optical fiber communications has been reviewed. The field has not yet been fully exploited; a number of interesting scientific and technical tasks remain open for research in the future.

For those readers who want to investigate the field deeper, principal review papers are mentioned in the References. These are, in a chronological order, (Yamamoto and Kimura, 1981), (Favre et al., 1981), (Okoshi and Kikuchi, 1981), (Okoshi, 1982), (Hooper et al., 1983), (Okoshi, 1984), (Okoshi, 1986), (Okoshi, 1987). A textbook covering this field is also available (Okoshi and Kikuchi, 1987).

REFERENCES

Cotter, D. (1983). Stimulated Brillouin scattering in
    monomode optical fiber. Jour. Opt. Commun., Vol. 4,
    No. 1, pp. 10-19.

Favre, F. and LeGuen, D. (1980). High frequency stability
    of laser diode for heterodyne communication systems.
    Electron. Lett., Vol. 16, pp. 709-710.

Favre, F. et al. (1981). Progress towards heterodyne-type
    single-mode fiber communication system. IEEE Jour.
    Quantum Electron., Vol. QE-17, No. 6, pp. 897-906.

Hooper, R. C. et al. (1983). Progress in monomode trans-
    mission techniques in the United Kingdom. IEEE/OSA
    Jour. of Lightwave Tech., Vol. LT-1, No. 4, pp. 596-
    611.

Minemura, K. (1987). Progress in coherent optical trans-
    mission research. Opt. Fiber Commun. Conf. (OFC/IOOC
    '87), Reno, Nevada, Jan. 19-22, Paper No. WF-1.

Mochizuki, K. (1986). Application of coherent optical com-
    munications technology to undersea communications, A
    panel talk at Panel Session C10: "Coherent Optical
    Communications," in the First Optoelectronics Conf.
    (OEC '86), Tokyo.

Okoshi, T. (1979). Feasibility study of frequency-division
    multiplexing optical fiber communication systems using
    optical heterodyne or homodyne schemes (in Japanese).
    Pap. Tech. Group, IECE Japan, No. OQE78-139, February
    27.

Okoshi, T. and Kikuchi, K. (1980). Frequency stabilization
    of semiconductor lasers for heterodyne-type optical
    communication schemes. Electron Lett., Vol. 16, No. 4,
    pp. 179-181.

Okoshi, T. and Kikuchi, K. (1981). Heterodyne-type optical
    fiber communications. Jour. Opt. Comm. Vol. 2, No. 3,
    pp. 82-88.

Okoshi, T., et al. (1981). Computation of bit-error rate
    of various heterodyne and coherent-type optical com-
    munication schemes. Jour. Opt. Comm., Vol. 2, No. 3,
    pp. 89-96.

Okoshi, T. (1982). Heterodyne and coherent optical fiber communications: Recent progress (Invited). IEEE Trans. Microwave Theory Tech., Vol. MTT-30, No. 8, pp. 1138-1149.

Okoshi, T. (1984). Recent progress in heterodyne/coherent optical fiber communications. IEEE/OSA Jour. of Lightwave Technol., Vol. LT-2, No. 4, pp. 341-346.

Okoshi, T. (1986). Ultimate performance of heterodyne/coherent optical fiber communications. IEEE/OSA Jour. Lightwave Technol., Vol. LT-4, No. 10, pp. 1556-1562.

Okoshi, T. (1987). Recent advances in coherent optical fiber communication systems. IEEE/OSA Jour. Lightwave Technol., Vol. LT-5, No.1, pp. 44-52.

Okoshi, T. and Kikuchi, K. (1987). Coherent Optical Fiber Communications. KTK(Tokyo)/Reidel(Dordrecht).

Stolen, R. H. (1980). Nonlinearity in fiber transmission. Proc. IEEE, Vol. 68, No. 10, pp. 1232-1236.

Yamamoto, Y. (1979). Study on optical digital modulation-demodulation systems (in Japanese). Pap. Tech. Group IECE Japan, No. OQE 79-144, October 25.

Yamamoto, Y. (1980). Receiver performance evaluation of various digital optical modulation-demodulation systems. in the $0.5-10 \mu m$-wavelength region. IEEE Jour. Quantum Electron., Vol. QE-16, No. 11, pp. 1251-1259.

Yamamoto, Y. and Kimura, T. (1981). Coherent optical fiber transmission systems. IEEE Jour. Quantum Electron., Vol. QE-17, No. 6, pp. 919-935.

5

Present and future of research on wave propagation

R. K. CRANE

INTRODUCTION

This tutorial considers the progress in radio science in the areas of interest to URSI Commission F, wave propagation and remote sensing, reported for the period between the last General Assembly and this assembly. Directions for future research are also considered. The material used in the preparation of this tutorial was compiled from the submissions of the member committees of URSI for the preparation of the section of the Review of Radio Science for Commission F [Crane, 1987]. Over nine hundred reference citations were received: journal articles, books, reports and symposia presentations. The large number of citations illustrates the vigor of the commission and the extent of the research program to be reviewed. Some of the references were to applications of remote sensing but not to the science or techniques employed for remote sensing. URSI is the appropriate forum for the discussion of radio techniques for sensing and communications but the applications of the fruits of radio science are best considered by other unions.

The majority of the citations were on remote sensing. A wide range of techniques and applications were in evidence. Satellite based radars and radiometers were used for the study of the land surface, ocean and atmosphere. Ground based radars were used for the investigation of both clear and inclement weather and an airborne Doppler radar was employed for storm research. New Doppler and multiple-polarization weather radars are being deployed by various national weather services. A new class of upward looking, passive vertical temperature and water vapor sounders in combination with a network of wind profiling radars has been proposed for the replacement of the radiosonde network. A wide range of active and passive remote and in situ sensing tools have been employed for the study of sea ice and of snow on land surfaces. Although listed by application, the work of interest to Commission F was in the science of the remote measurement technique; in the interaction between the electromagnetic waves and the surface to be sensed or the volume to be probed.

Radio propagation studies in support of communication system design focused on the continued exploration of the use of millimeter wave frequencies, on the development of the transmission loss models needed for the design of land mobile satellite communications systems, on the investigation of frequency selective multipath fading to obtain an improvement in the performance of terrestrial digital radio links and on the characterization of the roles of ducting, scattering by turbulence, and scattering by rain in the production of interference between systems operating in the same frequency band. The European community finished a major multination study of the effects of rain on satellite communications and now has an interference study in progress. The Olympus Propagation Experiment planned by the European Space Agency promises to provide a wealth of data for the development of the higher frequency bands.

Theoretical studies of wave propagation through random media and of scattering by rough surfaces were reported by a number of researchers. New models were developed for the prediction of rain effects and improved models were prepared for the estimation of attenuation by atmospheric water vapor.

This report is on recent research in the fields of interest to URSI Commission F. It considers the general topic of wave propagation through non-ionized media. The presentation is organized by phenomena. The state of our knowledge in predicting the effects of each phenomenon on communications or remote sensing is explored. Attention is drawn to the compilation of nearly 300 journal citations in the current edition of the Review of Radio Science [Crane, 1987]. That review constitutes the primary listing of references for this tutorial.

PROPAGATION THROUGH THE CLEAR ATMOSPHERE

Gaseous absorption   The complex index of refraction describes the macroscopic interaction between electromagnetic waves and a medium. Quantum mechanical models are required for the microscopic description of the interaction between the waves and individual molecules or collections of molecules and to bridge the gap between the microscopic and macroscopic descriptions. A succession of approximate models have been used for this purpose to calculate the specific absorption and radio refractivity of the constituent gases of the atmosphere. The semi-empirical quantum mechanical model for oxygen absorption has been quite successful but the estimation of water vapor effects has been a problem for more than 40 years. An empirical correction to the semi-empirical model is required to reconcile measurements and model predictions [Bohlander et. al, 1985].

About a decade ago, Russian workers suggested that the discrepancies between measurements and model predictions could be due to two water vapor molecule linkages called dimers. Since that time, several experimenters reported the observation of anomalous water vapor absorption relative to the empirical models for high humidity values at temperatures near $0^\circ C$. Detailed experimental studies of water vapor absorption were conducted under a wide range of physical conditions to quantify the discrepancies. Several studies reported during the past triennium have again identified the cause for the lack of agreement as dimers while others show no apparent conflict between the empirical models and observations.

Microwave remote measurement of atmospheric water vapor using emission, absorption or phase delay techniques may be in error for conditions of high humidity, the conditions of most interest for the development of clouds and rain. The model predictions needed for the determination of communications link reliability in the window regions of the millimeter wave spectrum may also be in error. More research is needed to reconcile satisfactorily the models and measurements.

Refraction   Spatial variations in the real part of the refractive index produce bending, focusing, ducting, multipath and, if they occur on a small enough scale, reflection and scattering phenomena. Studies of ducting and multipath phenomena are in progress in a number of countries to address reliability issues associated with the high bit error probabilities that accompany frequency selective fading on digital radio links. The atmospheric conditions that give rise to frequency selective fading also can produce high level field strengths over long distances and thereby increase the potential for interference between spatially separated systems sharing the same frequency band. The European community is now conducting a major study to provide the experimental data needed to develop interference models.

Extended period experimental observations of the statistical variations of the transmission loss between terminals in a communications service or between systems in an interference problem are required to provide the information needed for model development. Supporting meteorological data are also necessary for model development. The latter are critical for interference studies because it is not practical to make measurements for every possible geometrical configuration of system locations, antenna sizes and pointing directions and terrain features. Most potential interference problems will have to be addressed using computer simulations based on valid models.

Terrestrial line-of-sight links may become important tools for the remote sensing of meteorological phenomena. At the higher frequencies, multi-frequency phase dispersion measurements could provide information about path averaged water vapor and liquid water. Path attenuation measurements may also yield important data on precipitation along the path. For precipitation, measurements at small attenuation values will be required to obtain a reasonable dynamic range. Means to observe and compensate for focusing will be required on such a path before an acceptable measurement technique is perfected.

Turbulence   An interest in the development of communications systems at millimeter wave frequencies has sparked a renewed interest in scintillation due to atmospheric turbulence along the propagation path. Measurements are now being reported at frequencies in the transmission windows up through 250 GHz.

Scattering by turbulence provides a means to sense atmospheric motions. Wind profiling radars are being developed to provide continuous wind measurements. Statistical analyses of scintillation on closely spaced terrestrial paths ma;y also bc uscd to sense air motion and, if the fluctuations due to a single constituent such as water vapor can be isolated, the transport of the constituent.

Scattering by turbulence provides the propagation mechanism employed by troposcatter systems for communication over paths of 150 to 300 km. Although interest in the use of troposcatter for communications has waned over the past two decades, the scatter mechanism provides a way to couple unwanted or interfering signals between systems. To characterize the interference problem at the higher frequencies, troposcatter measurements are now being made at frequencies above 10 GHz.

PROPAGATION THROUGH AND SCATTERING BY HYDROMETEORS

Rain   Over four decade ago rain was identified as the major culprit in questions of communications link reliability at frequencies above 5 GHz. Since then, the problems associated with propagation through rain have continued to attract the attention of the propagation community. We are still working to characterize the rain medium and to predict the effects of rain on link performance. In the early days we were content to investigate the physics of the interaction between electromagnetic waves and rain. Now we are attempting to provide information useful in characterizing the limitations imposed by propagation through rain on the bit error probabilities for wideband digital transmission, on the use of different coding schemes for the reduction in the errors produced by moderate attenuations, on the use of frequency reuse by orthogonal dual polarizations, on the use of site diversity , and on up-link power control and antenna switching for down-link power control on satellite systems.

Statistical studies of the occurrence probabilities of rain effects exceeding specified limits have been conducted in a number of different geographical locations and climate regions. A strong interest exists in the development of adequate meteorologically based statistical models for the occurrence of rain effects because the intended application is always to a location lacking a long time series of measurements.

The information needed to characterize statistically the temporal and spatial structure of rain for the prediction of propagation effects and for the development of adequate meteorological models is obtained from weather radars. The weather radar is a remote sensing device that has gained acceptance by the meteorological community and the public at large. In many locations, television broadcasts of weather forecasts include displays from a weather radar if rain is present. New Doppler weather radar systems are now being deployed by a number of national meteorological services to provide improved local forecasts.

The development of new tools for the remote sensing of precipitation is still underway. Multiparameter radars are being used to study the detailed cloud physics of the development of convective storms. Multi-frequency terrestrial line-of-sight links are being constructed for the remote sensing of the average precipitation accumulation over an area. New satellite systems are being developed for the remote sensing of precipitation from space. These systems are for use over water in regions devoid of any other sensor system. Theoretical studies suggest the potential use of active and passive sensors on low orbiting satellites for over land observation of convective rain.

The new tools for remote sensing require the use of adequate meteorological models for the state of the precipitation process to effect the inversion of the measurements. Most sensing systems do not measure the parameter of interest. For the remote measurement of rain rate at the surface, radars observe the backscattering from raindrops, radiometers observe the emission from the drops and line-of-sight systems can observe the path integrated attenuation or phase dispersion. In each case, the parameter that is measured is not linearly related to the rain rate and is often not at the surface. The statistical characterization of the rain drop size distribution has been the preoccupation of the weather radar community for nearly three decades because the drop size distribution sets the inversion relationship between the observations and the parameter to be estimated. For sensing systems with large fields of view, the inversion problem is complicated because the parameter to be observed varies significantly within a resolution element of the observing system. Insufficient information is available to correct for the nonlinearity between the observed and desired parameter and the variation of the parameter within a resolution cell. Model based methods are required to solve this problem.

We are just beginning the development of adequate inversion procedures for the remote sensing of precipitation. The development process requires a close collaboration between radio scientists and meteorologists to develop models that take both the interaction between the electromagnetic waves and the medium and the spatial and temporal structure of that medium into account. The promise of a multiparameter radar system is the direct determination of a sufficient number of parameters within a resolution volume to statistically characterize the drop size distribution within the volume. If however the rain rate increases between the height of the measurement and the ground due to meteorological processes, the estimate of rain rate at the surface is still in error.

Snow   Snow present an interesting problem both for remote measurements and for the estimation of interference effects. The characterization of the particle size, type, orientation and dielectric properties distributions is not as precise as for rain. Winter storms are usually not as deep as summer storms and with a rapid decrease in reflectivity with height, snow is difficult to observe over a large area with a radar system.

Snow has attracted little interest from the communications community because the attenuation produced by snow at frequencies that have been used for communications is so small as to be negligible for system design. Large attenuation events have been reported for snow conditions but the effect is usually attributed to the occurrence of melting snow on the path or to wet snow on the antenna.

A possible problem for communications system designers is interference by scattering from snow at the higher frequencies. One model for interference by precipitation scatter suggests that the problem can be neglected at higher frequencies because the attenuation along the path reduces the scattered signal below any reasonable level for interference. Snow can still scatter the signal but the attendant attenuation expected for rain is absent. To date the highest troposcattter signals reported for a 165 km path at Ku band have been with snow along the path.

Clouds   Clouds of moderate liquid water content can produce measurable to severe attenuation events for ground to space links at millimeter waves. Clouds contaminate the fields of view of most infrared and visible wavelength remote sensing systems. Unfortunately, no satisfactory remote sensing system currently exist for the measurement of the liquid water content and spatial extent of clouds. The information needed for communication system design does not exist. Recent work on the development of millimeter wave radars holds some promise for the measurement of cloud properties.

EMISSION AND SCATTERING BY A ROUGH SURFACES AND PROPAGATION THROUGH RANDOM MEDIA

Sea   The remote sensing of the sea surface was a subject which received significant attention during the past triennium. Active and passive measurements from satellites and aircraft were available for analysis. A number of studies were reported on the use of synthetic aperture radar techniques. Ground based HF radars provided information on surface winds and currents over wide areas. Analyses of scattering and emission by waves on the ocean surface continued to keep pace with the available measurements.

In large measure, the focus of the sea surface sensing work has shifted from the development of measurement tools to the applications of these tools to oceanographic observations and the interpretation of the observations in terms of ocean processes. Some studies did consider the measurement accuracies achievable by remote sensing techniques.

Land   Soil moisture measurements were the focus of a number of land surface remote sensing studies. Reflections from the terrain were studied to determine the levels of multipath fading and delay spread to be expected for communications links in the land mobile satellite service. Applications to that service were the source of a number of theoretical studies and simulations during the past triennium.

Vegetation   Vegetation affects the emissivity and reflectivity of the land surface as viewed from above. It also attenuates the signal. Studies were made to determine the effects of vegetation on the remote sensing of soil moisture and to characterize signatures of different crops in different states or stress. Several tree and forest attenuation measurements were conducted to obtain design data for land mobile communications systems. Experimental measurements were made with mechanical analogs of single plants and groups of plants to determine the statistics of the scattered and transmitted fields with known distributions of leaf shape, size, orientation and number density. Computer simulations were made in conjunction with the measurements.

Buildings   Studies of transmission though buildings and scattering by buildings were made to provide information for the design of cellular mobile radio communications systems. Measurements and computer simulations of the probability of finding clear lines-of-sight through urban areas were also made.

Ice and Snow   A number of theoretical studies of the scattering properties of ice and snow on the land surface or over water were reported during the past triennium. Both volume scattering and rough surface models were employed. Advances were made in the description of the scattering properties of individual grains and voids in the snow volume, in describing the macroscopic properties of the snow medium and in the multiple scattering treatment of propagation through the snow volume.

REFERENCES

Bohlander, R.A., McMillan, R.W.and Gallagher, J.J. [1985]: Atmospheric Effects on Near-Millimeter-Wave Propagation, Proc IEEE, 73, 49-50.

Crane, R. K. [1987]: Commission F - Wave Propagation and Remote Sensing, in Review of Radio Science 1983 - 1986, Ed. G. Hyde, International Union of Radio Science, Brussels, Belgium.

6

Aspects of ionospheric physics relevant to radio propagation

H. RISHBETH

ABSTRACT

Following a brief historical review in Section I of how ionospheric science developed, the principles of ionospheric modelling are described in Section II. Section III deals with some problems of the quiet ionosphere, and Section IV with the high-latitude ionosphere and its interaction with the magnetosphere. Section V summarizes relevant aspects of solar-terrestrial physics; Section VI is a brief conclusion.

I.   INTRODUCTION : AN HISTORICAL VIEW

The ionosphere may be defined as the part of the atmosphere in which free electrons are sufficiently numerous to influence the propagation of radio waves. It is conventionally divided into "regions" - D, E, F - with boundaries between regions at 90 and 150 km, though these boundaries should be considered to be governed by physics, not by statute. Within each region exist one or more "layers" of ionization, distinguished if necessary by symbols (e.g. F1, F2, Es). The limits of the ionosphere are not well-defined, but for practical purposes the lower boundary may be taken to lie at about 60 km, and the upper limit where protons become dominant over the oxygen ions of the F2-layer, several hundred kilometres higher up.

The "discovery" of the ionosphere is usually attributed to Marconi, with his transatlantic radio transmission in 1901, though the idea of a current-carrying "conducting layer" had been mooted in the nineteenth century to explain the daily geomagnetic variations. Kennelly and Heaviside suggested that the reflection of radio waves might be due to free electric charges, and in 1903 Taylor proposed that these charges might be produced by ionizing solar radiation. Systematic exploration of the ionosphere only began with the experiments of Appleton & Barnett (1925) and Breit & Tuve (1925), which identified distinct stratifications or "layers". The term "ionosphere" was coined by Watson-Watt in 1926 (Gardiner, 1969).

The practical importance of the ionosphere lies in its effect on radio-wave propagation. Up till the mid-1950's, it was probably true to say that most ionospheric scientists were knowledgeable both in the physics of the ionosphere and in its propagation aspects. The physical quantities, such as electron production and loss rates, were investigated by analysing ionosonde and propagation data. For this purpose, diurnal variations and eclipse effects seemed especially useful (deceptively so, as it later turned out). Much of today's basic ionospheric theory had been developed by the mid-1950's, some of it by the mid-1930's.

There were perhaps two main factors that changed the science. The first was increased specialization, brought on by the explosion of knowledge that started in the IGY, largely through the use of satellites, rockets and incoherent scatter radars. The second was the growth of aeronomy, the science of the whole upper atmosphere. The ionosphere plays a key part in aeronomy, largely because its electrical properties and chemical activity give the ionization an importance out of all proportion to its relative abundance (even in the F2-layer, only about 0.1% of the air is ionized). Because the ionization is easier to observe and measure than the neutral gas, it serves as a valuable "tracer" for many upper atmosphere processes. It was the growth of knowledge about the neutral atmosphere that led to the great advances in ionospheric theory from the mid-1950's onwards. Then, as the wider field of solar-terrestrial physics was developed, ionospheric science fell into its place, along with the science of the magnetosphere, the interplanetary medium and the Sun.

Progress in this direction, however, seemed to reduce contact between the "propagation" and "scientific" sides of ionospheric science, and so the "engineering" and "physics" sections of the community tended to diverge. A contributory factor was the change in the nature of ionospheric communications, the traffic being increasingly handled by "adaptive" systems that measure and respond to the vagaries of the ionosphere. Nevertheless, demands still exist for long-term "predictions" and short-term "forecasts" of ionospheric conditions, and there must still be plenty of scope for bringing the advances of ionospheric science and solar-terrestrial physics to bear on practical needs. This paper briefly explores some ways in which progress might be made. It does not deal with the actual physics of radio propagation as such, nor with radio communication systems, but selects some relevant aspects of present-day ionospheric physics.

II. MODELLING THE IONOSPHERE

Ionospheric science started with measurements of the ionosphere's physical properties. Because of the great spatial and temporal variability that was discovered, it became necessary to describe the ionosphere systematically - i.e. its variations with height, place and time. This has mainly been accomplished by swept-frequency radar sounders - ionosondes - of which only a few operated in the 1930's, but about 150 during the IGY in 1957-58 and still about 100 in the mid-1980's. Many other techniques are now available, ground-based and space-based.

From both "propagation" and "scientific" points of view, the most important characteristic of the ionosphere is the electron density vs height profile, N(h). Propagation conditions largely depend on the peak electron densities NmE, NmF1 and NmF2 of the layers (related to the "ordinary critical frequencies" or penetration frequencies at vertical incidence, foE, foF1, foF2).

By way of illustration, Fig. 1 shows the variation of monthly mean critical frequencies at noon, at Slough (England) over fifty years; the seasonal and sunspot cycle effects are obvious. The daytime E and F1 layers follow quite well the formula

$$\sqrt{Nm} \propto fo \propto (1 + aR).(\cos \chi)^n \qquad (2.0)$$

where χ is the solar zenith angle, a is a constant, R is the sunspot number, and the index n lies in the range 0.2-0.35. The normal D layer (though it does not possess a critical frequency) also varies quite regularly with solar zenith angle, but the F2 layer does not.

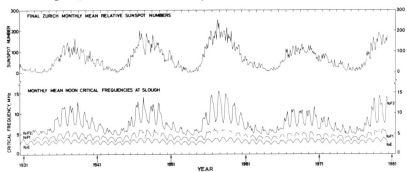

Fig. 1. Monthly mean noon critical frequencies of the F2, F1 and E layers at Slough, U.K. (Lat. 52N), compared to the mean Zürich sunspot number, 1931-1981. Copyright: S.E.R.C. Rutherford Appleton Laboratory, U.K.

A proper study of the ionosphere must both describe these variations and explain them. The first step is to build up a "database", an organized collection of data. Example are the archives of "vertical incidence sounding data" - ionospheric critical frequencies and related parameters - in the ionospheric World Data Centres. To progress beyond this stage, "modelling" is often employed. The purposes and problems of ionospheric models have been discussed by Nisbet (1974) and several other authors.

There are two basic kinds of models, either of which may be useful for predicting and forecasting: "empirical" and "physical". An "empirical" or "descriptive" model comprises mathematical formulas or computational algorithms, with numerical constants that are determined empirically. These models seek to represent the observed values of physical parameters as closely as possible, but are not necessarily based on physics. The model of N(h) profiles between the E and F2 layers, developed by Bradley & Dudeney (1973) is of this type. A more comprehensive example is the International Reference Ionosphere (IRI) (Rawer, 1981). Another example is the MSIS global model of the neutral thermosphere (Hedin, 1983), constructed from formulas that have been fitted to worldwide data yielded by Mass Spectrometers aboard satellites and Incoherent Scatter radars at several places on the ground. The MSIS model is based on formulas that represent the global temperature distribution. Physical principles, such as the equation of state and the barometric law, are then used to compute gas densities and other parameters.

A "physical" model, usually "theoretical" or "computational" in nature, simulates the behaviour of the ionospheric electron density by solving the physical equations that govern it: equations of state, continuity, motion, energy. Apart from the task of setting up and solving the equations, the main problems are to include all the important processes; obtain good values of the various parameters; and to set appropriate boundary conditions. More sophisticated models solve equations, not only for the density N but also for the velocity V and temperature T of the electrons and ions; and more comprehensive models treat also the corresponding equations for the neutral air, in order to take account of its important interactions with the electrons and ions. For any kind of particle, these equations can be written schematically as follows:

Continuity equation   (Conservation of mass)

$\partial N/\partial t$ =   {Production} - {Loss} - {Transport}   (2.1)

Force equation   (Conservation of momentum)

dV/dt = {Driving force} - {Drag} - {Advection} *   (2.2)

Heat equation   (Conservation of energy)

$\partial T/\partial t$ =   {Heating} - {Cooling} - {Conduction}   (2.3)

 * Advection: transport of momentum by viscosity, etc

The first real "physical" model of the ionosphere was
the "Chapman layer", an ionized layer that is produced by
solar photoionization and subject to loss by recombination
(Chapman, 1931).   The theory gives rise to the formula
(2.0) mentioned above, with the index n = 0.25 for a
"classical" Chapman layer. But the model is also used in
an "empirical" way, in which the formula (2.0) is assumed
and a value of n is fitted to data for daily, seasonal or
latitudinal variations of fo. The values of n are merely
descriptive; they often differ from 0.25, and physical
reasons for the difference may or may not be known.

III. THE QUIET IONOSPHERE AT MIDDLE AND LOW LATITUDES

Although the behaviour of the quiet ionosphere is
quite well known and reasonably well understood, there
are many phenomena that bear on radio propagation, but are
not fully explained. The list of such phenomena is long
(though only a few can be discussed here), and includes:

> D-region absorption, links with meteorology
> Sporadic E
> Large-scale anomalies of the F-layer
> Spread F and related irregular structure
> Day-to-day variability
> Place-to-place variability
> Waves & their effects

A. The Problem of Variability.   Like most natural
systems, the ionosphere (even under magnetically quiet
conditions) shows a great deal of variability. The normal
E-layer is the least variable, being accurately
predictable on an hourly and daily basis, at least for a
given level of solar activity. Variability is greatest
for the F2 layer, being more than 10% in foF2 from day to

day, with place-to-place variations that have some
systematic features. By way of illustration, Fig. 2
shows the variation of foF2 throughout complete years at
sunspot minimum and sunspot maximum. The seasonal and
storm effects are obvious, and many other interesting
features can be found.

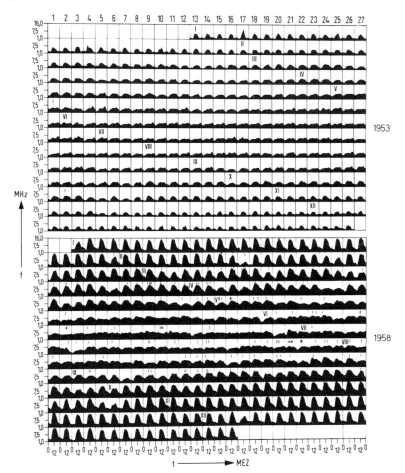

Täglicher Verlauf der Grenzfrequenz foF2 in Lindau/Harz
I Mögel-Dellinger-Effekt

Fig. 2. "Lindau Mountain Diagrams", showing the
critical frequencies foF2 at Lindau (Lat 52 N) throughout
the sunspot minimum year 1953 and the sunspot maximum year
1958. Each box represents 00-24 LT and 1 - 16 MHz; the
data are arranged in lines of 27 days in order to display
any recurrences attributable to the solar 27-day rotation
period. Copyright: Max-Planck-Institut für Aeronomie,
Katlenburg-Lindau, German Federal Republic.

B. **F-layer modelling.** Some of the most challenging problems, to which present-day ionospheric physics might be applied, are in the field of F2-layer behaviour. Physical modelling essentialy means solving the continuity equation (2.1), which for the F2 layer can be written as

$$\partial N/\partial t \quad = \quad q \quad - \quad \beta N \quad - \quad \text{div } (NV) \tag{3.0}$$

where q and β are the rates of production and loss, and V is the transport velocity due to winds, electric fields and plasma diffusion. As is well known, solutions of (3.0) are in general complicated to obtain, and difficult to "calibrate" in terms of real data. Perhaps the main reason is that the physical parameters and coefficients in the equation are not easily measured, and most are perhaps known only to within an accuracy of say 50%. Production, loss and diffusion coefficients of course depend on the neutral air density, and therefore any useful physical solution of (3.0) requires such knowledge, which can only come from a reliable "empirical" or "physical" model of the neutral upper atmosphere (thereby introducing a host of problems that are not really ionospheric).

One practical possibility is to use "semi-empirical" models, which simplify the equations so that they become computationally fast to solve, and are then used to generate numerical coefficients for an empirical model (e.g. Rush et al. 1984, Anderson et al., 1985). For mid-latitudes, the F2-layer "servo" model (Rishbeth, 1967) represents the layer as a quasi-equilibrium system in which the height hm, responding to vertical drift, tends at any time towards an "equilibrium" value, primarily determined by diffusion and loss rates but modified by vertical drift W(t); the peak density Nm depends on the ambient production and loss rates at height hm. The "servo" equations may be written

$$dhm/dt \quad = \quad D \ (he - hm) \quad + \quad W(t) \tag{3.1}$$

$$dNm/dt \quad = \quad q(hm) - c \ \beta \ (hm) \tag{3.2}$$

where D is related to the diffusion coefficient and c is a numerical constant that can be estimated from theory. Such a computationally simple approach, if properly calibrated, might in principle provide a useful tool for F2-layer propagation calculations and even predictions.

C. **Sporadic E.** Sporadic E, or Es, is important from the point of view of radio propagation. The high Es critical frequencies that occur, coupled with the sporadic and unpredictable nature that gives "sporadic E" its name, can present severe problems to radio communications. It is thus worth considering what physical understanding exists of Es, in the hope that understanding might eventually lead to some degree of predictability.

Twelve standard classifications of sporadic E, based on the appearance of ionograms, have been defined (Piggott & Rawer, 1972). These classifications are morphological rather than physical, and their usefulness may be limited. Physically there may be only three main types:

(1) Equatorial Es, a plasma instability caused by the high electron drift velocity (i.e. the large current density) in the daytime equatorial electrojet. Its physics seems reasonably well understood (e.g. Farley, 1985), though interesting questions remain, especially regarding the "counter-electrojet" which is associated with some other ionospheric phenomena. Since it is fairly regular and its properties are well known, the equatorial electrojet does not present any great practical problem.

(2) Particle Es, caused by precipitation of auroral particles, which has serious effects on propagation at high latitudes, but might best be treated as part of the general problem of the auroral oval and its associated phenomena (Sec. IV.A).

(3) Mid-latitude Es is thought to be produced by wind shears, i.e. small-scale gradients of wind velocity. The sheared wind interacts with the geomagnetic field in such a way as to compress ions into thin layers, typically 1 km thick but ^{2} 100 km in horizontal extent. Theory shows that, to exist for minutes or hours as observed, these layers must comprise long-lived ions rather than the ordinary gaseous ionospheric ions. It is generally accepted that these ions are metallic, formed by ionization of trace constituents such as Fe, Mg and Ca. Some forms of these layers are associated with semi-diurnal tidal winds, which are fairly regular and might have some predictability. Others are formed by smaller-scale winds, possibly by gravity waves, and for practical purposes are truly "sporadic". So, even where the likelihood of Es is known statistically, reliable predictions in detail seem beyond reach at present.

D. **Waves and their Effects**. Only one point is made here, but it is an important one: whenever ionospheric effects are attributed to "waves" of any sort, one should consider exactly what processes, induced by the wave, are actually operating in the continuity equation (2.1).

IV.   THE HIGH-LATITUDE IONOSPHERE

A.   Auroral Oval and Associated Phenomena.   **In low** and middle latitudes (up to magnetic latitudes of about 60 deg, depending on conditions) the geomagnetic field lines are closed; but at higher latitudes they are linked to the magnetosphere, the tenuous region around the Earth that is permeated by the geomagnetic field. The high latitude ionosphere is profoundly affected by auroral particles and electric currents, that originate in the magnetosphere. The particles and currents mainly enter the atmosphere in the "auroral ovals", rings 2000-3000 km in diameter that surround the magnetic poles. In turn, the magnetosphere is influenced by the solar wind - the stream of charged particles emitted by the Sun - and by the interplanetary magnetic field.

Many geophysical phenomena are caused by auroral particle precipitation. Hartz & Brice (1967) defined two types of precipitation, the intense impulsive "splashes" of relatively soft keV particles associated with substorms, and the more persistent background "drizzle" of harder MeV particles at rather lower latitudes. Some of these phenomena affect radio wave propagation, so it is useful to describe them. Examples given by Hartz & Brice include:

"SPLASH EVENTS"   "DRIZZLE EVENTS"

Discrete auroral forms   Steady diffuse aurora
Auroral absorption events   Slowly-varying absorption
Auroral sporadic E   Low diffuse sporadic E
Rapidly fading VHF scatter   VHF forward scatter
Impulsive Pi pulsations   Continuous Pc pulsations
Bursts of VLF auroral hiss   Quasi-constant VLF chorus
Short bursts of keV electrons   Sustained electron fluxes
Bremsstrahlung X ray bursts   Long-duration hard X rays
Negative magnetic bay

Since many of these phenomena influence radio wave propagation, any scheme that organizes the data on their occurrence is potentially useful.

B.   Modelling the Auroral Oval.   **In principle it** should be possible to represent the position of the auroral oval in terms of geographic, geomagnetic and solar-terrestrial parameters, either "empirically" or "physically" (in the sense explained in Sec. II). Then the positions of ionospheric features could be related to the geometry of the auroral oval. This has been done for auroral absorption events by Foppiano & Bradley (1985). If a way could then be found to link the model parameters to observable solar or interplanetary quantities (with any necessary time delay), this might form the basis of a model of high latitude propagation characteristics. Such a model might have potentialities for either short-term "forecasting" or long-term "prediction", if the key parameters could be either "forecasted" or "predicted". In principle, this might lead to a practical scheme.

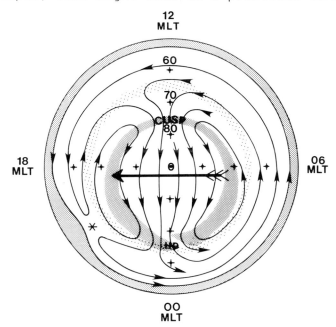

Fig. 3. **The auroral oval and associated phenomena.** View of the north polar region, centred on the magnetic pole, in a non-rotating frame of reference. Magnetic midnight (00 MLT) is at the bottom; magnetic noon (12 MLT) at the top; the outer boundary is at 50 deg magnetic latitude; magnetic latitudes 80, 70, 60 deg are marked by crosses. The outer ring represents the plasmasphere (mid-latitude ionosphere) which co-rotates with the Earth. The stippled ring represents the approximate location of the hard "drizzle" precipitation. The inner shaded ring is the auroral oval, showing the approximate locations of the "cusp" in the noon sector and the Harang discontinuity (HD) near midnight. The fine lines represent typical flow lines of the plasma convection pattern (some curves are left incomplete to reduce congestion); note the stagnation point at 20 MLT. The diagram corresponds roughly to moderate magnetic activity (K = 3) with a southward Bz component of the interplanetary magnetic field. Copyright: S.E.R.C. Rutherford Appleton Laboratory, U.K.

To this end, Fig. 3 is a schematic plan of the northern auroral oval and features linked to it. The oval can in fact be quite well represented by a circular ring, the centre of which is offset by a few degrees, on the midnight side, from the magnetic dip pole. Both the mean latitude (M) and offset (S) depend on the level of geomagnetic activity, which in turn depends on solar activity and the configuration of the interplanetary magnetic field. This leads to the following equations for the geomagnetic latitude of the oval, in terms of local "geomagnetic time" GT (in hours) and the magnetic disturbance K-figure, of the type

$$L = M - S \cos (GT \times 15) \qquad (deg) \qquad (4.0)$$

Poleward boundary of the oval (geomag. lat. Lp):

$$Mp = 73.7 - 0.1 \ K \quad (deg) \qquad (4.1)$$

$$Sp = 3.9 - 0.45 \ K \quad (deg) \qquad (4.2)$$

Equatorward boundary of the oval (geomag. lat. Le):

$$Me = 71.8 - 0.9 \ K \quad (deg) \qquad (4.3)$$

$$Se = 4.6 + 0.15 \ K \quad (deg) \qquad (4.4)$$

The precise formulation of these equations is a matter for debate, since the parameters involved (magnetic latitude, magnetic activity and time) can be defined in different ways. If the "magnetic K-figure" is taken to be the high latitude "quarter-hourly index" Q, the equations fit quite well the northern hemisphere auroral data of Feldstein & Starkov (1967). For present purposes, it is not crucial which parameter is best; the answer may well be different for different aspects. The point is that a numerical model of the auroral oval should be realizable.

C. **Maintenance of the Polar Ionosphere.** **From** early days in ionospheric physics, it was a puzzle as to how the ionosphere was maintained throughout the polar winter in the absence of solar ionizing radiation. Since the polar caps (within the auroral ovals) are not on the whole locations of strong particle precipitation, and the nighttime sky provides only weak sources of ionizing radiation, a possible solution to the puzzle might lie in terms of horizontal transport of ionization.

It is now known that the magnetic linkage between the solar wind, as it sweeps past the Earth, and the high-latitude geomagnetic field leads to a fast day-to-night drift of ionospheric plasma across the polar cap (Knudsen, 1974). This appears - at least in principle - to supply plasma to the nightside of the polar ionosphere, although this supply may be finely balanced because the rapid drift also leads to an increase in the decay rate of the plasma. This day-to-night drift, sketched in Fig. 3, is part of a large-scale circulation or convection of plasma, which takes the form of closed loops. At sub-auroral latitudes, the convection pattern also seems to account for the trough, a region of low plasma density. In particular, the combination of the Earth's rotation and the convection velocity causes plasma to become almost stationary with respect to the Sun, in the evening sector. If this stagnation point (shown at 20 GT in Fig. 3) is in darkness, the plasma density can decay to very low levels, as has been observed.

V. STORMS AND THEIR IONOSPHERIC EFFECTS

A.   Summary of Solar-Terrestrial Relations.   **From** early days (Appleton & Ingram, 1935) it was known that magnetic storms are accompanied by "ionospheric storms" or "disturbances", in which the electron density distribution becomes modified in ways that can seriously affect radio communications. Magnetic storms are in turn related to solar disturbances, notably major solar flares. The complicated chain of cause and effect involves the solar wind. A brief summary may be given as follows.

The Sun emits a stream of charged particles, the solar wind, which travels to the Earth at speeds of 300-1000 km/s with a transit time of 1.5-3 days. The solar wind originates in coronal holes. It is greatly intensified and accelerated by solar disturbances, especially by the plasma streams emitted by solar flares. Thus magnetic disturbances are, in general, associated with variations in the solar wind, which show some degree of 27-day periodicity due to the Sun's rotation. The occurrence of major flares facing the Earth also has some 27-day periodicity, but only because flares tend to be associated with active sunspot groups which rotate with the Sun. The extent to which geomagnetic disturbances ensue depends on the orientation of the interplanetary magnetic field, IMF.

Flares emit: (a) some visible light; (b) hard X-rays (0.1-1 nm) that produce greatly enhanced ionization in the lower ionosphere, thus causing "sudden ionospheric disturbances" - fadeouts, VLF propagation anomalies, etc, starting at the onset of the visual flare and lasting up to 1 hour; (c) MeV electrons that arrive a few minutes later, and (d) MeV protons taking about 1 hour to reach the Earth, these causing absorption events in high latitudes; (e) energetic plasma streams that travel to the Earth as strengthenings of the solar wind. The complex terrestrial effects of these plasma streams include:

1. Magnetospheric compression; enhanced energy input leading to frequent and violent auroral substorms; the storm-time ring current.

2. Intensified activity in the auroral ovals, with expansion towards lower latitudes.

3. Geomagnetic disturbances; the storm-time decrease due to the magnetospheric "ring current"; magnetic bays due to auroral currents.

4. Ionospheric storm effects at all latitudes; short-term and long-term D-layer effects; F-layer negative and positive storms.

    B. Problems of Ionospheric Storms. **The ionospheric** effects are widespread, unpredictable in detail, but are most important from the point of view of radio propagation. Some examples are clearly visible in Fig. 2. However, some patterns of behaviour are known, and might in principle provide some predictability, if only on an empirical basis. To some extent the storm effects are believed to be "transmitted" from auroral latitudes towards lower latitudes (e.g. Davies, 1974); but just how much that occurs, and whether it could form the basis of short-term forecasting, is still a matter for investigation. Of interest, and some practical value, is the fact that some geomagnetic effects of the IMF's orientation can be monitored at the ground.

    A complete physical description of ionospheric storms is still remote, even in a purely theoretical way. Perhaps the ideal to aim for is a theoretical ionospheric model, well calibrated against real data, with appropriate modelling of ionosphere/solar wind/IMF links; and coupled with IMF and solar wind monitoring, and solar flare predictions. Whether that is realizable is conjectural.

    VI. CONCLUSION

    This essay has outlined some of the ionospheric physics that affects propagation and predictions; but the subject is a vast one, and only a tiny fraction of its literature has been cited here. A useful survey of the topic of short-term predictions of disturbances, with many references, may be found in CCIR Report 727-1, 1982.

Several questions remain. One concerns the prediction
of median quiet-day parameters of the ionosphere, which
are useful for long-term planning of services. Since
most quiet-day ionospheric phenomena are quite well
documented (even if not fully explained) such predictions
are largely a matter of predicting solar activity, and
choosing the best parameters to relate solar activity to
ionospheric parameters (such as IF2, Smith 1968). A more
difficult question concerns the short-term prediction or
forecasting of deviations from mean behaviour. In theory,
these are unnecessary if all communications systems are
"adaptive", and capable of responding in real time to
changing conditions. In practice, this is not so; and
even if it were so, there would still be value in
forecasting the likelihood of ionospheric disturbance,
which again requires the monitoring of solar and
interplanetary conditions. Furthermore, there might still
be an operational need for forecasting the progress of
ionospheric storms, once started. This is a challenging
field, and many ideas have been advanced, some based on
the idea of storm effects being "transmitted" from higher
to lower latitudes.

Many lines of further study suggest themselves for the
future. For the present, the fact is that ionospheric
monitoring and prediction services are kept in being -
because they continue to serve a useful purpose.

ACKNOWLEDGEMENTS

Acknowledgement for the diagrams is made to Rutherford
Appleton Laboratory, U.K., and the Max-Planck-Institut für
Aeronomie, Katlenburg-Lindau, F.R.G. Some passages of the
text are taken by permission from Chapter 5, "Basic
Physics of the Ionosphere" in Radio Wave Propagation
(editors M P M Hall & L W Barclay), IEE/Peter Peregrinus,
Stevenage, U.K. The assistance of Professor A L Cullen,
OBE, FRS in the production of the final text is gratefully
acknowledged.

REFERENCES

Anderson D N, Mendillo M & Herniter B (1985). AFGRL
    TR85-0254.
Appleton E V & Barnett M A F (1925). Nature 115 **333-334.**
Appleton E V & Ingram I J (1935). Nature 136 **548-549.**
Breit G & Tuve M A (1925). Nature 116 **357.**
Bradley P A & Dudeney J R (1973). J. Atmos. Terr. Phys.
    35 **2131-2146.**
Chapman S (1931). Proc. Phys. Soc. Lond. 43 **26-45.**
Davies K (1974). Planet. Space Sci. 22 **237-253.**
Farley D T (1985). J. Atmos. Terr. Phys. 47 **719-744.**

Feldstein Y I & Starkov G V (1967).   Planet. Space Sci.
    15 **209-229**.
Foppiano A J & Bradley P A (1985). J. Atmos. Terr. Phys.
    47 **663-674**.
Gardiner G W (1969).   Nature   224 **1096.**
Hartz T R & Brice N M (1967).   Planet. Space Sci. 15
    301-329.
Hedin A E (1983).   J. Geophys. Res. 88   **10170-10188.**
Knudsen W C   (1974).   J. Geophys. Res. 79 **1046-1055.**
Nisbet J S (1975).   Atmospheres of Earth and the Planets,
    Reidel, Dordrecht, 245-258.
Piggott W R & Rawer K (1972). UAG-23, WDC-A, Boulder, Colo.
Rawer K (1981). UAG-82, WDC-A, Boulder, Colo.
Rishbeth H (1967).   J Atmos Terr Phys. 29 **225-238.**
Rush C M, PoKempner M, Anderson D N, Perry J, Stewart F G
    & Reasoner R (1984).   Radio Sci. 19 **1083-1097.**
Smith P A (1968).   J. Atmos. Terr. Phys. 30 **177-185.**

7

Present and future trends in research in waves in plasmas

W. S. KURTH AND S. D. SHAWHAN

ABSTRACT

This tutorial presents a summary of the present status of
plasma wave research, primarily in space plasmas, and an overview
of research to be expected in the next several years. Some treatment of new work in the laboratory which is applicable to waves in
space plasmas is also discussed.

INTRODUCTION

Research in waves in plasmas is approaching a plateau of
having completed a basic phase of exploration of waves in space
plasmas and achieving a sound foundation upon which much more
detailed studies can be based. Fundamental plasma wave spectra
have been measured in and near the magnetospheres of Earth,
Jupiter, Saturn, and Uranus, in the solar wind interaction region
surrounding Venus, and within the comets Giacobini-Zinner and
Halley. Plasma wave studies at the Earth have progressed to a
state of reasonable qualitative (and some quantitative) agreement
with theory and have, therefore, made the exploration of other
regimes in the solar system a much more tractable problem. Plasma
wave research at the Earth is beginning to be carried out as an
integral part of the physics of the magnetosphere and ionosphere
and not simply a study of isolated phenomena. There is a definite
synergism between problems under study in naturally occurring
plasmas and well-conceived laboratory experiments.

Plasma wave research in the vicinity of the Earth, or
geospace, is revealing a steadily decreasing number of unique, new
phenomena with the trend moving towards a greater understanding of
now-familiar plasma wave emissions (Shawhan, 1985). Fig. 1 (from
Shawhan, 1979) depicts the wide range of magnetospheric plasma
waves and their usual locations. There are very few plasma wave
emissions which cannot boast a long list of theoretical studies
designed to explain the generation and propagation of the waves.
Numerical simulations are being utilized with ever-increasing frequency to explore non-linear or quasi-linear instabilities and to

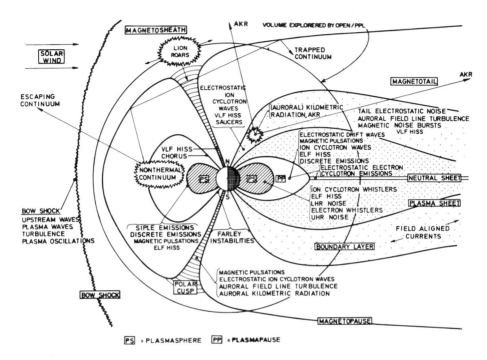

Fig. 1. A schematic drawing showing the wide variety of plasma wave and radio phenomena existing in the earth's magnetosphere (Shawhan, 1979).

understand the complex consequences of rapidly growing modes on the underlying plasma distribution functions. In fact, the quality of the theoretical studies of several problems has progressed to the point that more complete observations of wave vectors and plasma distribution functions are required to validate the theories.

The most positive and forward-looking developments in plasma wave research in the last few years have been the realization of the importance of folding details of the wave-particle inter- actions into the basic physical processes going on in naturally- occurring plasmas. It is this focus that will guide the future of plasma wave research. The increase in knowledge of plasma waves in geospace and other plasma regimes in the solar system should naturally lead to a better understanding of astrophysical plasmas and the generation of radio emissions therein.

The future of research in waves in plasmas will be the focus of this tutorial. While as stated above the field is largely beyond its exploratory stage, there is still significant exploration to be undertaken within the next several years. The major exploratory efforts include extending the study of plasma waves to planets such as Neptune and Mars as well as to small bodies; revisiting comets and studying the solar wind interaction near asteroids. The broadest movement in the next several years will be towards in-depth knowledge of the generation of currently known instabilities, and more importantly, the relevance of those instabilities to the physics of geospace. The latter point involves understanding further the role of plasma waves in the management of the magnetospheric energy budget and in the acceleration and loss of plasmas. A relatively new era of active experiments is dawning which involves the launching of waves in the ionosphere from a wave injection facility as well as understanding the generation of instabilities from beams of charged particles or the release of chemical tracers in a plasma. Finally, it would be incorrect to omit ongoing and future efforts in the laboratory, since these experiments lay a crucial foundation to our understanding of waves in space plasmas. While the boundary conditions are drastically different in a laboratory chamber, critical parameters of the plasma can be controlled in a much more efficient manner and a more complete set of diagnostic data can be obtained.

EXPLORATORY RESEARCH IN PLASMA WAVES

Waves in Planetary Environments. Plasma wave observations of varying degrees of detail and coverage have been obtained at Venus, Earth, Jupiter, Saturn, and Uranus. The probes to Mercury and Mars did not carry sensors to provide plasma wave information on the solar wind interaction with those planets. (It is incredible that we still do not have a consensus on the existence of an intrinsic magnetic field at Mars.) There are no currently-planned missions to Pluto. We look forward to advancing our exploration of plasma waves near the planets primarily at Mars and Neptune. The next Soviet mission to Mars, Phobos, will carry sophisticated plasma wave instrumentation and will hopefully provide at least a cursory view of plasma waves in the vicinity of that planet. Unfortunately, Phobos will likely not spend any extensive time close enough to the planet to provide an extensive data set. The Mars Aeronomy Observer, still in the very early planning stages, should provide detailed observations of the solar wind interactions with either a Martian ionosphere or magnetosphere, including the full spectrum of plasma waves.

Perhaps the most exciting event imminent in planetary plasma wave exploration lies in the August 1989 Voyager encounter of Neptune. In addition to providing the first observations of plasma waves in the vicinity of the last gas giant, a unique high

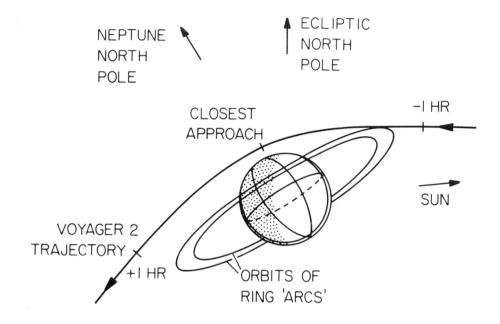

NEPTUNE NORTH POLE

ECLIPTIC NORTH POLE

−I HR

CLOSEST APPROACH

SUN

VOYAGER 2 TRAJECTORY

+I HR

ORBITS OF RING 'ARCS'

TRAJECTORY PLANE

Fig. 2. The trajectory of Voyager 2 will take it very close to the north pole of Neptune, providing a unique opportunity to observe plasma waves at high latitudes in a non-terrestrial magnetosphere.

latitude, low altitude trajectory shown in Fig. 2 provides the possibility of studying a realm of plasma wave phenomena which have not been observed except at the Earth. The trajectory carries Voyager 2 to a latitude of about 76 deg. and to within about 4700 km of the cloud-tops of Neptune. The close approach followed by a solar occultation by Neptune would seem to assure measurements of the solar wind interaction with Neptune, even in the case of no intrinsic magnetic field (solar wind/ionosphere interaction). Based on similarities of Neptune to Uranus, it would seem more likely that Neptune has a well-formed magnetosphere and a nearly-aligned dipole configuration would lead to a passage though high magnetic latitudes, possibly even through the cusp, auroral oval, and polar cap. The large tilt and offset of the Uranian dipole suggests an aligned dipole is not a foregone conclusion; however, even large tilts would provide fertile regions for plasma wave studies in a non-terrestrial environment.

Should the Voyager spacecraft pass though or close to auroral field lines at low altitudes, the possibility of observing auroral hiss exists which would enable ray-tracing studies to determine the location of the field-aligned plasma flows exciting the emission. While the relative velocity of the spacecraft will be high, it is possible that evidence for ion-cyclotron emissions associated with high latitude ion conics would be available. At the Earth, this trajectory would be ideal for the study of the auroral kilometric radiation source region, hence, we look forward to a close-up or even in situ look at a likely Neptunian radio emission source region.

The Voyager encounter of Neptune will provide not only plasma wave observations, but also in situ observations of the magnetic field, plasma, and energetic particle populations both at low altitudes as well as at large distances, perhaps on similar field lines. This correlative data set combined with ultraviolet surveys of the planet for auroral activity could result in a Dynamics Explorer-type of study of the Neptunian magnetosphere/ionosphere coupling problem. Further, the spacecraft passes through the equatorial plane at distances just beyond the known ring-arc material both inbound on the dayside as well as outbound at night, allowing day/night comparisons of the magnetic equatorial region (assuming a nearly aligned dipole) where critical wave-particle interactions should be taking place.

Plasma Waves in the Vicinity of Small Bodies. The pioneering work in research in plasma waves involved in the solar wind interaction with comets has already been done, based on the observations by International Cometary Explorer (ICE) at Giacobini-Zinner (Scarf et al., 1986), and Vegas 1 and 2 (Grard et al., 1986; Klimer et al., 1986), Sakigake and Suisei (Oya et al., 1986) at Halley in the past two years, not to mention the exploratory research afforded by the AMPTE artificial comet experiments (Gurnett et al., 1985). Still, the comet observations have only whet our appetites. As shown in Fig. 3, the ICE observations show an enormous interaction region associated with Giacobini-Zinner, stretching millions of kilometers beyond the shock-like feature otherwise denoting the solar wind interface (Scarf et al., 1986). Follow-on missions such as Comet Rendezvous and Asteroid Flyby (CRAF) offer the opportunity to study the temporal evolution of the comet/solar wind interaction, both on a time scale commensurate with the perihelion passage, as well as with short time scale variations in the solar wind such as the passage of sector boundaries.

CRAF and possibly Galileo will provide our first observations of the solar wind interaction with asteroids. While this interaction is likely to be quite benign, one might expect interesting wake effects to be produced in the downstream plasma

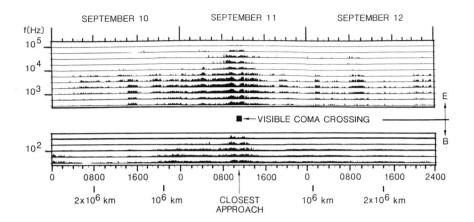

Fig. 3. An overview of the ICE plasma wave observations in the vicinity of comet Giacobini-Zinner. The solar wind-comet interaction region extends to extremely large distances as evidenced by the plasma wave turbulence. Reprinted by permission from Science, Scarf et al., Vol. 232, pp. 277-381, 18 April 1986, copyright 1986 by the AAAS.

cavity. The wake is likely to be turbulent and perhaps accentuated by broadband electrostatic modes. It is probable that the asteroids present a rather inert surface in terms of a source of plasma, but sputtering by solar wind plasmas could produce sufficient material so that pickup effects would be noted.

Plasma Waves in the Heliosphere. The eventual launch of Ulysses will provide the first survey of the solar wind at high solar latitudes and will enable us to compare waves and instabilities in that region with those we are familiar with close to the ecliptic plane. The evolution of the solar wind is dependent on variations with latitude; our lack of information at high latitudes prevents a complete understanding of the formation of the heliosphere.

The Voyager mission provides a unique opportunity to study the occurrence and characteristics of plasma wave phenomena in the distant heliosphere, and in the poorly understood interface region with the interstellar wind. Evidence from Voyager thus far points to a decrease in characteristic frequencies of the solar wind plasma in response to decreasing solar wind plasma densities and magnetic field strengths with increasing distance. Further, the overall level of plasma wave turbulence appears to fall off with distance; however, this effect has not been studied quantitatively to date. Voyager has provided data out to more than 30 AU and at large distances out of the ecliptic plane; we expect data to continue for the next several years from the Voyager Interstellar Mission (VIM). Estimates of the distance to the inner heliospheric shock range from 30 AU to beyond 100 AU, some of the smaller values based on an interpretation of the source of low frequency radio emissions observed by Voyager since 1983 (Kurth et al., 1987). It is anticipated that the Voyager investigations will provide plasma wave spectra and accompanying plasma and magnetic field information at the terminal shock and within the subsonic solar wind beyond (Fig. 4). In view of the large range of uncertainty of the conditions in the outer heliosphere and in the vicinity of the heliopause, we can look forward to some very interesting results from the Voyager plasma wave and associated investigations.

ADVANCED PLASMA WAVE STUDIES

<u>**Towards a Comprehensive Understanding of Geospace.**</u> Perhaps the introduction to this tutorial treated the current state of understanding of plasma waves in geospace unfairly. Several of our colleagues might argue that great strides have, indeed, been

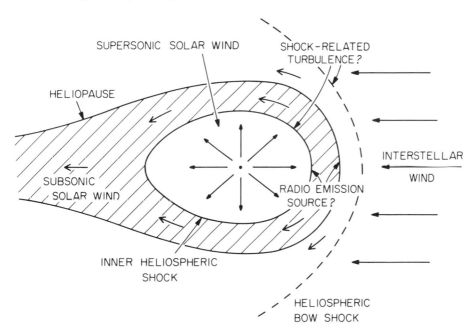

Fig. 4. A schematic representation of the heliosphere after Petelski (1981) showing the possible source locations of low-frequency radio emissions observed by Voyager and the various interaction surfaces which may be sites of plasma wave turbulence.

made in unfolding the interrelationships between plasma wave
turbulence and the energization and loss of plasma in the
magnetosphere (Anderson, 1983). We do not dispute these claims of
success.   Rather, we argue that such successes are required in a
more global view of the interaction of the magnetosphere with the
solar wind.   We can also cite cases where our understanding of
some processes involving plasma waves has actually required a
reassessment.   A prime example is the Belmont et al. (1983) conclusion that, contrary to the popular belief for some ten years,
electron cyclotron waves may not be the sole, or even primary,
precipitation agents for the diffuse aurora.

The International Solar Terrestrial Physics (ISTP) program is
designed to attack the deficiencies in our understanding of the
overriding processes governing the response of the magnetosphere
to variations in the solar wind input as well as a comprehensive
understanding of the mass and energy budget of the magnetosphere.
ISTP seeks to accomplish these goals through a battery of
simultaneous, multipoint observations by spacecraft situated in
the upstream solar wind (Wind/USA, SOHO/ESA) to monitor the solar
input; in the deep magnetotail (Geotail/ISAS) to study energy
storage, particle acceleration, and substorm processes; in polar
orbit (Polar/USA, Cluster/ESA) to study energy deposition into the
aurora as well as measure the ionospheric plasma flux and near
the equator (CRRES/USA, Cluster/Equatorial Science Phase) to study
the ring current and plasma sheet (Fig. 5).

ISTP provides the opportunity to fly a new class of plasma
wave receiver; e.g., the instrument on Polar seeks to capture the
full vectorial components of both the electric and magnetic components of waves.   Such measurements permit the unambiguous determination of wave mode through measurement of wave polarization and
wave vector direction as well as the calculation of such quantities as the Poynting flux (Shawhan, 1983).   Supporting plasma distribution functions with high temporal resolution, composition
determination, and particular attention to the elusive loss cone
will replace speculation with a solid observational basis for
theoretical studies.

Extraterrestrial Plasma Waves Revisited. The next decade
provides the opportunity to improve our understanding of plasma
physical processes at two crucial sites: the sun and Jupiter.
The astrophysical importance of these two plasma regimes is well
recognized.   SOHO, as a part of ISTP provides a unique opportunity
to couple enhanced observations of the sun with a coordinated set
of terrestrial-based monitors of the resulting response of the
magnetosphere. Ulysses, Wind and SOHO will all provide additional
data of the distant solar wind.   Solar Probe, a possible new
mission before the year 2000, could provide the first observations
of plasma and plasma wave phenomena to within 4.5 solar radii.

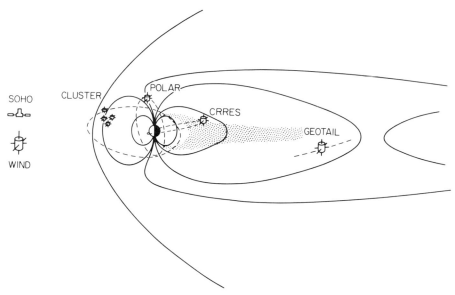

Fig. 5. The ISTP program will provide multipoint observations in geospace in order to study the fundamental processes of energy transport, storage, and deposition in the magnetosphere.

The Galileo orbiter at Jupiter provides an opportunity to obtain radio and in situ observations in and near the Jovian magnetosphere for several months and allows the study of temporal variations not possible with flyby missions (Fig. 6). Further, Galileo will be able to distinguish with much more confidence between local time, latitude, and radial distance effects as well as to understand the mechanisms of the clock-like periodicities in particle fluxes and radio emissions from the planet. In many instances, the instrumentation on Galileo is more sophisticated, by far, than the Voyager and Pioneer instrumentation which preceded it. This is especially true in the case of plasma wave instrumentation, which for the first time, includes a magnetic search coil antenna as part of the Galileo sensor complement.

Theory and Simulation. The study of plasma physics in space requires a strong infrastructure of theoretical research to support the observations described above. This is particularly true for advancements in the understanding of plasma waves in geospace, for as the observations increase in resolution and completeness, the theories must follow with increased fidelity. Part of the increase in theoretical scope of recent years is the recognition of the importance of the thermal plasma and the increased attention to the thermalization of free energy in the

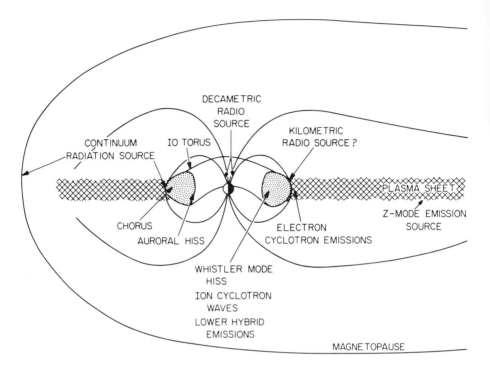

Fig. 6.   The Jovian magnetosphere is the location of a number of interesting plasma wave and radio phenomena. The rapidly rotating magnetic field and rich plasma source at Io create unique contrasts with the terrestrial magnetosphere for study by Galileo.

plasma distribution function. Often, the warm plasma effects are associated with non-linear or quasi-linear saturation mechanisms. While an analytic approach is necessary to gain insight into the physics of the situation, increasing reliance on numerical simulations is required to fully understand the processes.

Continued progress in the theoretical research on plasma waves looks promising with the continued support for programs such as the Solar Terrestrial Theory Program. While some of this effort deals with large-scale modeling, a significant portion of this and other programs' resources is focussed on such problems as heavy-ion heating by ion-cyclotron waves (Ashour-Abdalla and Okuda, 1984) and investigations into broadband electrostatic noise associated with field-aligned ion streaming (Dusenbery, 1986) as well as other plasma wave phenomena. Other theoretical work is crucial in understanding the generation of planetary radio emissions, such as auroral kilometric radiation (Winglee and Pritchett, 1986). It will be crucial for ISTP to be supported by a strong theory program to gain the maximum benefit from the coordinated study of geospace.

ACTIVE EXPERIMENTS

Active experiments involve the stimulation of natural plasmas by one or more artificial (man-made) stimuli in order to study the response of the plasma in a more-or-less controlled manner. In a very real sense, this type of experiment brings an advantage of laboratory plasma physics into the large scale plasmas found in geospace. The intent is to decouple the source of the perturbations from the plasma system in which the experiment is done so that the stimulus can be varied at will in time, intensity, or location.

Until recently, active experiments have been limited primarily to rocket experiments and, in fact, the rocket program continues to play a large role in active plasma experiments. As depicted in Fig. 7, the Spacelab missions have opened a whole new opportunity for active experiments which enable a hands-on approach to performing the experiments (Shawhan et al., 1983). It could be argued that Spacelab's most important contribution to plasma physics is the provision for extended active experiment opportunities. A decade from now, similar experiments may be performed from the Space Station.

Wave Injection Experiments. Wave injection experiments are perhaps the active experiments which are most intimately associated with plasma wave research since they allow the introduction of waves into a plasma under controlled conditions, allowing for the study of propagation characteristics and such facets as triggering or stimulated precipitation. The Waves in Space Plasma (WISP) experiment being developed for Space Plasma Lab for flight sometime after 1993 is the primary wave injection experiment to be considered here. WISP comprises a VLF and an HF section covering the frequency range from below 1 kHz to 30 MHz. The antenna is planned to be a tubular dipole extended from the shuttle to a maximum tip-to-tip length of 300 m. Major objectives include studying the propagation of VLF waves, the measurement of phase velocities, and the stimulation of other VLF emissions or perhaps electron precipitation. The HF portion of the experiment allows the excitation of waves at frequencies ranging from the vicinity of the ionospheric plasma frequency (a few MHz) to well above the characteristic frequencies of the plasma. A crucial part of the WISP investigation is the capability for bistatic measurements with the use of a Recoverable Plasma Diagnostics Package (RPDP) orbiting within a few hundred km of the shuttle. A Soviet low frequency wave injection experiment,

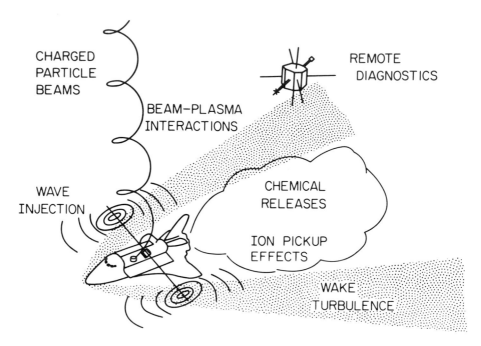

Fig. 7. Spacelab has created new opportunities for active experiments in the ionosphere including the injection of charged particle beams, low to high frequency waves, and chemical tracers.
In addition, studies of wake phenomena are possible. The use of a
remote diagnostics package is crucial to the completion of these
active experiments.

Aktivny-1K is scheduled for launch in 1987 and includes the use of
a subsatellite as well.

   Charged Particle Beam Experiments.   One of the more puzzling
problems in plasma physics is the propagation of a beam of
electrons or ions through and the interaction of the beam with a
plasma.   Charged particle beams can also be used as virtual
antennas for the emission of waves.   An example of the plasma
waves generated in such a beam plasma interaction is given in Fig.
8 which shows the funnel-shaped VLF hiss generated by an electron
beam on Spacelab 2 (Gurnett et al., 1986).   A number of such
experiments have been flown on rocket missions, a fast-pulse electron beam was flown on STS-3 (Shawhan et al., 1984) and Spacelab 2
(Gurnett et al., 1986) and experiments which produced ion as well
as electron beams were flown on Spacelab 1 (Obayashi et al., 1984;
Beghin et al., 1984).   One of the Spacelab 1 experiments, Space
Experiments with Particle Accelerators (SEPAC), is being prepared

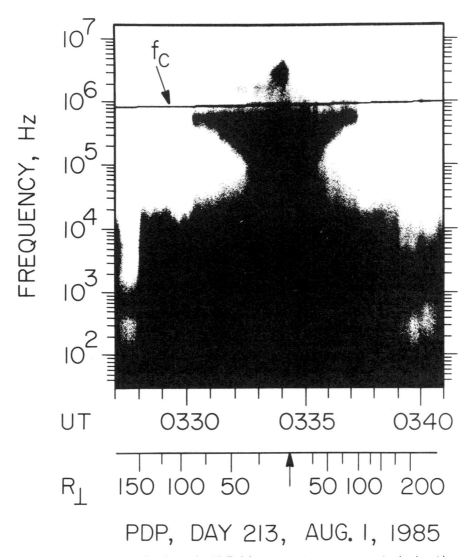

Fig. 8. This funnel-shaped VLF hiss event was generated by the injection of a 1-keV electron beam on Spacelab 2, showing the relevance of beam experiments to the study of auroral phenomena such as auroral hiss (Gurnett et al., 1986).

to fly on Atlas 1 in 1992 and Space Plasma Lab at a later time. The obvious application of the SEPAC and related experiments are to study the interaction of the beams with a plasma to gain insight into auroral processes, as an example. Remote diagnostics are required and will be provided by a co-orbiting diagnostics package RPDP on Space Plasma Lab. The Soviet Union is planning an electron accelerator experiment in 1989, Apex, which will also use a daughter satellite for diagnostics.

The rocket program continues to promote the study of electron beam interactions. Charge III (1989) will be the third in a series of electron beam experiments and will include a tether used as an antenna. ECHO-7 will carry an electron accelerator and attempts will be made to monitor wave generation both on the ground and from satellites.

Chemical Release Experiments. Chemical releases have taken place in the terrestrial magnetosphere and ionosphere for several years, primarily through the rocket program. It hasn't been until relatively recently, however, that in situ plasma wave observations were included in such experiments. The most notable series of experiments are the Active Magnetospheric Particle Tracer Experiment (AMPTE) artificial comet and magnetotail releases (Gurnett et al., 1985, Haüsler et al., 1985). Shuttle flights provide an excellent opportunity to study the effects of chemical releases in the ionosphere, using the effluents from the shuttle's engines or simply the outgassing products as the contaminating agents (Pickett et al., 1985). Recent reports from Spacelab 2 show evidence for intense, broadband electrostatic waves and for ion pickup distributions which may be intimately involved with the wave instability (Gurnett et al., 1987; Paterson and Frank, 1987). Future work in this area is expected with the launching of the Combined Release and Radiation Effects Satellite (CRRES) and with additional experiments performed from the shuttle, for example, as part of the Space Plasma Lab experiments.

Wake Studies. Recent shuttle flights, including STS-3 and Spacelab 2 have provided a whole new regime in which to study the flow of plasma around an obstacle. The shuttle is a large object with respect to virtually all relevant plasma scale sizes, including the thermal ion gyroradius. Early reports from these experiments indicate general agreement with predictions from the theory of plasma expansion. Plasma turbulence is evidently important in the physics of the flows as evidenced by the apparent localization of intense turbulence at the gradient in the density on the boundary of the wake (Murphy et al., 1986; Stone et al., 1986).

Future experiments will be undertaken as part of Space Plasma Lab utilizing a new generation of diagnostics instrumentation being developed for that program, the RPDP, which will include a plasma wave receiver with a digital waveform capture capability that simultaneously captures signals from three orthogonal electric and three orthogonal magnetic sensors in a bandwidth up to 30 kHz, theoretically enabling the identification of the mode of the turbulence. Also anticipated is the first flight of the Electrodynamic Tether, enabling studies of plasma flows past the tethered object, with the ability to control the potential of the object over a very wide range as well as the ability to use the tether as an antenna to launch waves in the ULF/ELF/VLF range.

LABORATORY

Laboratory plasma experiments play an important role in plasma wave research. Though the chamber sizes are limited and the range of plasma parameters achievable is somewhat restricted, laboratory research can attack a wide range of problems which are difficult to study in space plasmas. In some cases, lab work can investigate issues in much greater depth because conditions can be created in which various phenomena thrive and parametric studies can be carried out by methodically varying one or more parameters. Recently, many issues have been addressed in the lab which are prominent problems in space plasma wave research; work in many of these areas promises to make important contributions to our understanding of plasma waves in the next several years.

One plasma wave mode central to several space plasma physics problems is the electrostatic ion cyclotron (EIC) wave and its relationship to ion conics observed in the auroral ionosphere (Klumpar, 1979). Recent work in the lab by Cartier et al. (1986) has shown the perpendicular heating of ions and subsequent transfer of perpendicular energy into both parallel thermal energy and parallel drift energy in the presence of a diverging magnetic field. In another laboratory effort, the effect of ion-neutral and electron-neutral collisions has been investigated in order to understand how low in the ionosphere the EIC modes can become unstable (Suszcynsky et al., 1986). It was determined that ion-neutral collisions always tend to stabilize the mode, however, the electron-neutral collisions can destabilize the waves. By varying the mixture of different species of ions, simulating the hydrogen-oxygen mixture in the ionosphere, Suszcynsky et al. show the waves can be unstable as low as 130 km, in agreement with both theory and observations.

Work continues in the lab on double layers. Recently, Alport et al. (1986) have studied the occurrence of EIC waves associated with a double layer in an inhomogeneous magnetic field. While no single mechanism adequately accounts for the observed EIC waves, there are certain characteristics consistent with both a current-driven mechanism and a perpendicular electric field model.

In work closely related to the study of plasma flows and wake structures near the shuttle, considerable effort is continuing to

be put into the study of plasma flows around obstacles in
laboratory plasmas (c.f., Wright et al., 1985, D'Angelo and
Merlino, 1986). The lab effort is especially important for wake
studies, in spite of new data from experiments such as those
carried out during STS-3 (Murphy et al., 1986) and more recently,
Spacelab 2, because it is difficult to unambiguously sort out the
variations of plasma parameters as well as geometrical factors
such as magnetic field direction in order to fully understand the
relationships.

Laboratory studies are even delving into such issues as magnetic reconnection. Experiments with magnetic field configurations having neutral points and variations which force plasma to
flow in various directions relative to the neutral point have met
with some measure of success although there is some debate as to
the direct applicability to geophysical processes (Stenzel et al.,
1981). Nevertheless, it is definitely of interest to further
pursue such experiments, particularly from the point of view of
understanding the role plasma waves might have in the physics near
the neutral point.

Finally, laboratory studies can even be used in the investigation of radio emission processes, such as the nonlinear production
of electromagnetic radiation from electron plasma oscillations
(Whelan and Stenzel, 1985).

SUMMARY

While a complete synopsis of the progress in plasma wave
research was not given herein, it is clear that the field has
matured to a great extent. Some exploratory work needs to be
completed, mainly in the area of extending our observations to new
regimes in the solar system as permitted by current and upcoming
planetary missions. The field of research in waves in plasmas,
however, is most likely to benefit from the incorporation of an
understanding of wave-particle interactions into the overall study
of plasma and magnetospheric physics. It is essential that, while
we continue to learn more about the generation of isolated plasma
wave modes, we also seek to evaluate the relevance of those waves
to the underlying physics of geospace and other plasma regimes in
the solar system.

It is also important that we continue to move in the
direction of applying our in situ observations and accompanying
theories to the study of astrophysical plasmas, which are too
distant to hope for in situ observations in the foreseeable future
and generally more energetic, requiring some care in the
extrapolation process. A key ingredient in the application of
plasma physics based on solar system observations to the
astrophysical case is a detailed understanding of remote sensing

techniques and how the remote observations can be used to deduce the characteristics of the astrophysical plasma. Radio emissions are central to this issue, implying the importance of understanding completely the generation of planetary and solar radio emissions.

Spectrometry is also important, however, to the extrapolation of solar system plasma physics to astrophysical problems. It is crucial to know how to interpret spectra of astrophysical objects so as to gain the maximum amount of reliable information on the plasmas in and around the sources of interest. The landmark test case linking remote observations with familiar in situ plasma observations was the UV observations of the Io torus in the Jovian magnetosphere (Broadfoot et al., 1979) and the subsequent in situ observations of the plasmas inferred from the UV spectrum (Bridge et al., 1979).

ACKNOWLEDGEMENTS

The authors wish to acknowledge useful discussions with N. D'Angelo. The research at the University of Iowa was supported by NASA through Grant NGL 16-001-043 and through Contract 957723 with the Jet Propulsion Laboratory.

REFERENCES

Alport, M.J., Cartier, S.L., and Merlino, R.L. (1986). Laboratory observations of ion cyclotron waves associated with a double layer in an inhomogeneous magnetic field. J. Geophys. Res. 91, 1599-1608.

Anderson, R.R. (1983). Plasma waves in planetary magnetospheres. Rev. Geophys. Space Phys. 21, 474-494.

Ashour-Abdalla, M., and Okuda, H. (1984). Turbulent heating of heavy ions on auroral field lines. J. Geophys. Res. 89, 2235-2250.

Beghin, C., Lebreton, J.P., Maehlum, B.N., Troim, J., Ingsoy, P., and Michau, J.L. (1984). Phenomena induced by charged particle beams. Science. 225, 188-191.

Belmont, G., Fontaine, D., and Canu, P. (1983). Are equatorial electron cyclotron waves responsible for the diffuse auroral electron precipitation? J. Geophys. Res. 88, 9163-9170.

Bridge, H.S., Belcher, J.W., Lazarus, A.J., Sullivan, J.D., McNutt, R.L., Bagenal, F., Scudder, J.D., Sittler, E.C., Siscoe, G.L., Vasyliunas, V.M., Goertz, C.K., and Yeates, C.M. (1979). Plasma observations near Jupiter: Initial results from Voyager 1. Science. 204, 987-991.

Broadfoot, A.L., Belton, M.J.S., Takacs, P.Z., Sandel, B.R.,
    Shemansky, D.E., Holberg, J.B., Ajello, J.M., Atreya, S.K.,
    Donahue, T.M., Moos, H.W., Bertaux, J.L., Blamont, J.E.,
    Strobel, D.F., McConnell, J.C., Dalgarno, A., Goody, R.,
    McElroy, M.B. (1979). Science. 204, 979-982.

Cartier, S.L., D'Angelo, N., and Merlino, R.L. (1986). A
    laboratory study of ion energization by EIC waves and
    subsequent upstreaming along diverging magnetic field lines.
    J. Geophys. Res. 91, 8025-8033.

D'Angelo, N, and Merlino, R.L. (1986). The effect of a magnetic
    field on wake potential structures. IEEE Trans. Plasma Sci.
    PS-14, 609-610.

Dusenbery, P.B. (1986). Generation of broadband noise in the
    magnetotail by the beam acoustic instability. J. Geophys.
    Res. 91, 12,005-12,016.

Grard, R., Pedersen, A., Trotignon, J.-G., Beghin, C., Mogilevsky,
    M., Mikhailov, Y., Molchanov, O., Formisano, V. (1986).
    Nature. 321, 290-291.

Gurnett, D.A., Anderson, R.R., Häusler, B., Haerendel, G., Bauer,
    O.H., Treumann, R.A., Koons, H.C., Holzworth, R.H., Lühr, H.
    (1985). Plasma waves associated with the AMPTE artificial
    comet. Geophys. Res. Lett. 12, 851-854.

Gurnett, D.A., Kurth, W.S., Steinberg, J.T., Banks, P.M., Bush,
    R.I., and Raitt, W.J. (1986). Whistler-mode radiation from
    the Spacelab 2 electron beam. Geophys. Res. Lett. 13, 225-
    228.

Gurnett, D.A., Kurth, W.S., Steinberg, J.T., and Shawhan, S.D.
    (1987). Plasma wave turbulence produced by the shuttle:
    Results from the PDP free flight. Science. Submitted for
    publication.

Häusler, B., Woolliscroft, L.J., Anderson, R.R., Gurnett, D.A.,
    Holzworth, R.H., Koons, H.C., Bauer, O.H., Haerendel, G.,
    Treumann, R.A., Christiansen, P.J., Darbyshire, A.G., Gough,
    M.P., Jones, S.R., Norris, A.J., Lühr, H., and Klöcker, N.
    (1986). Plasma waves observed by the IRM and UKS spacecraft
    during the AMPTE solar wind lithium releases: Overview. J.
    Geophys. Res. 91, 1283-1299.

Klimov, S., Savin, S., Aleksevich, Ya., Avanesova, G., Balebanov, V., Balikhin, M., Galeev, A., Gribov, B., Nozdrachev, M., Smirnov, V., Sokolov, A., Vaisberg, V., Oberc, P., Krawczyk, Z., Grzedzielski, S., Juchiewicz, J., Nowak, K., Orlowski, D., Parfianovich, B., Wozniak, D., Zbyszynski, Z., Voita, Ya., and Triska, P. (1986). Extremely-low-frequency plasma waves in the environment of comet Halley. Nature. 321, 292-293.

Klumpar, D.M. (1979). Transversely accelerated ions: An ionospheric source of hot magnetospheric ions. J. Geophys. Res. 84, 4229-4237.

Kurth, W.S., Gurnett, D.A., Scarf, F.L., and Poynter, R.L. (1987). Long-period dynamic spectrograms of low-frequency inter-planetary radio emissions. Geophys. Res. Lett. 14, 49-52.

Murphy, G., Pickett, J., D'Angelo, N., and Kurth, W.S. (1986). Measurements of plasma parameters in the vicinity of the space shuttle. Planet. Space Sci. 34, 993-1004.

Obayashi, T., Kawashima, N., Kuriki, K., Nagatomo, M., Ninomiya, K., Sasaki, S., Yanagisawa, M., Kudo, I., Ejiri, M., Roberts, W.T., Chappell, C.R., Reasoner, D.L., Burch, J.L., Taylor, W.L., Banks, P.M, Williamson, P.R., and Garriott, O.K. (1984). Space experiments with particle accelerators. Science. 225, 195-196.

Oya, H., Morioka, A., Miyake, W., Smith, E.J., and Tsurutani, B.T. (1986). Discovery of cometary kilometric radiation and plasma waves at comet Halley. Nature. 321, 307-310.

Paterson, W.R., and Frank, L.A. (1987). Hot ion plasmas from the cloud of neutral gases surrounding the orbiter. Science. Submitted for publication.

Petelski, E.F. (1981). Generalized equations for the pressure transition and the stand-off distance of the solar wind terminating shock. J. Geophys. Res. 86, 4803-4806.

Pickett, J.S., Murphy, G.B., Kurth, W.S., Goertz, C.K., and Shawhan, S.D. (1985). Effects of chemical releases by the STS 3 orbiter on the ionosphere. J. Geophys. Res. 90, 3487-3497.

Scarf, F.L., Coroniti, F.V., Kennel, C.F., Gurnett, D.A., Ip, W.-H., and Smith, E.J. (1986). Plasma wave observations at comet Giacobini-Zinner. Science. 232, 277-381.

Shawhan, S.D. (1979). Magnetospheric plasma wave research 1975-1978. Rev. Geophys. Space Phys. 17, 705-724.

Shawhan, S.D. (1983). Estimation of wave vector characteristics. Adv. Space Res. 2, 31-41.

Shawhan, S.D. (1985). The menagerie of geospace plasma waves. Space Sci. Rev. 42, 257-274.

Shawhan, S.D., Burch, J.L, and Fredricks, R.L. (1983). Subsatellite studies of wave, plasma, and chemical injections from Spacelab. J. Spacecraft and Rockets. 20, 238-244.

Shawhan, S.D., Murphy, G.B., Banks, P.M., Williamson, P.R., and Raitt, W.J. (1984). Wave emissions from dc and modulated electron beams on STS 3. Radio Sci. 19, 471-486.

Stenzel, R.L., Gekelman, W. and Wild, N. (1981). Laboratory experiments on magnetic field line reconnection, in Physics of Auroral Arc Formation (ed. S.-I. Akasofu and J. R. Kan), pp. 398-407. American Geophysical Union, Washington.

Stone, N.H., Wright, Jr., K.H., Hwang, K.S., Samir, U., Murphy, G.B., and Shawhan, S.D. (1986). Further observations of space shuttle plasma-electrodynamic effects from OSS-1/STS-3. Geophys. Res. Lett. 13, 217-220.

Suszcynsky, D.M., Cartier, S.L., Merlino, R.L., and D'Angelo, N. (1986). A laboratory study of collisional electrostatic ion cyclotron waves. J. Geophys. Res. 91, 13,729-13,731.

Whelan, D.A., and Stenzel, R.L. (1985). Electromagnetic radiation and nonlinear energy flow in an electron beam-plasma system. Phys. Fluids. 28, 958-970.

Wright, K.H., Jr., Stone, N.H., and Samir, U. (1985). A study of plasma expansion phenomena in laboratory generated plasma wakes: Preliminary results. J. Plasma Phys. 33, 71-82.

8

Radio astronomy—new horizons

WM. J. WELCH

ABSTRACT

This is a brief review of recent progress in radio astronomy techniques with comments on the science that it enables. There have been important advances in receiver sensitivity, extension in operation to short wavelengths, improvements in angular resolution at all wavelengths, and developments in image processing. There have already been some successful experiments in space radio astronomy, and it is likely that the greatest gains in radio astronomical capability in the future will be made in space.

INTRODUCTION

Speaking to the General Assembly of the International Astronomical Union in 1961, Professor Jan Oort remarked that at that time the workshop of the astronomer was a very busy place. Indeed, there was an explosive development in astronomy at X–Ray, Radio, and Infrared wavelengths fostered by new instrumental capabilities. By 1961, Dutch and Australian radio astronomers had given us the first glimpse of the entire Milky Way through observations of the 21 cm line of neutral hydrogen, and the English and Australians, using their newly developed antenna arrays, had shown that the region beyond our galaxy was populated with numerous luminous radio sources.

In the quarter century since Professor Oort's comment, the pace of exciting new discoveries in radio astronomy has continued, supported by further technical advances. Accurate radio positions led to connections with classical astronomy; and some of the brightest radio sources were found to be galaxies and, within the Milky Way, supernova remnants. Then a new class of object was found, the Quasistellar Radio Source or Quasar. Improvements in receiver sensitivity and careful calibration studies in preparation for satellite communication led to the discovery of the 2.7K cosmic background radiation, a finding that has profoundly affected our view of the origin of the universe. A high time resolution array developed at Cambridge for the study of radio star scintillation found another unexpected new class of objects, the pulsars. The application of the results of laboratory microwave spectroscopy to radio astronomy led to the discovery of first the OH molecule and then a rich variety of polyatomic molecules in the interstellar medium. More recently, the study of emission from CO and other molecules has become an essential tool in the study of both star formation and galactic structure within our own and other galaxies. The invention of accurate atomic clocks permitted the development of disconnected Very Long Baseline Interferometry (VLBI), and the anticipated tiny bright cores of radio galaxies were found. The overall complexity of these sources, including superluminal motions, was unexpected. A profoundly important advance has been the VLBI study of proper motions in water maser sources that has expanded the useful distance of statistical parallax for distance measurement by two orders of magnitude. The big telescopes — the Bonn 100m, Arecibo, and others — and the newer arrays — the Cambridge 5km, the WSRT, and the VLA among the largest — with ever improving receivers, have steadily added to the picture: fainter source counts, more detailed structure of radio galaxies and quasars, neutral hydrogen in elliptical galaxies, steep spectrum galaxy cluster sources at long wavelengths, among other important discoveries. One of the most surprising developments, especially in the shortest wavelength work, has been the study of stars. Earlier it had seemed that stars, in some ways the most important astronomical objects, would never be much studied by radio astronomers, except for the Sun. Now it appears that our best hope of understanding how stars form is through millimeter wavelength studies with both single antennas and the new arrays. The envelopes and the evolution of the old red giant stars are also being investigated with these same tools.

In some ways, radio astronomers have behaved like gluttons, gobbling down the results produced by the new instruments with little digestion and rushing on to the construction of the next telescope in anticipation of the next feast. But that is an overstatement. Excellent survey work has been carried out and is now in progress. The surveys are part of the fundamental tool kit of the astronomer and are responsible for many of the important new discoveries and deeper understandings. An example is the first millisecond pulsar discovery, which would not have been found with the same techniques without the availability of several important surveys. On the technical side, the millisecond pulsars may become the time standards of the future.

OUR PRESENT HORIZON — THE NEW INSTRUMENTS

Recent technical developments have concentrated in four principal directions, receiver sensitivity, millimeter and submillimeter techniques, higher angular resolution systems, and computer image processing. Each step will have an important effect on the astronomical observations that can be performed. In some areas, fundamental limits are being approached, whereas in others many further developments may be possible.

1. Receiver sensitivity. The main effort here has been at short centimeter and millimeter wavelengths. At the longest wavelengths, $\lambda \geq 1m$,

receiver noise temperatures are already dominated by the brightness of the galaxy, and therefore only improvements in gain and phase stability are useful. In the past decade, Field Effect Transistors have brought down system temperatures of cm wavelength systems dramatically into the range of 30–50°K, at least for wavelengths greater than 3cm. Currently, the addition of the new High Electron Mobility Transistors promises to lower system temperatures by another factor of two or three in this range and by a factor of about five at 1cm. These developments will enable important observations of faint spectral lines. For example, until now the VLA has only been able to do limited high angular resolution studies of star formation regions in the important lines of ammonia at 1.3cm. The expected sensitivity improvement with the addition of HEMT amplifiers should offer a large number of additional line opportunities and produce a flood of important new results.

Further sensitivity improvements at cm wavelengths will have to come from more efficient use of the telescopes. Because of the cosmic background radiation, the emission from the atmosphere and residual noise pickup from the surroundings, it will be difficult to reduce system temperatures much below the 10–20°K levels now made possible by the HEMT amplifiers. Increasing sensitivity by a factor of two or more by enlarging the antenna collecting area over what is now in use does not seem practical. The Bonn 100m is about as large as it can be, based on the homology design, and still cover the cm wavelength band. Doubling the number of antennas in any of the larger arrays for a factor of two gain in instantaneous sensitivity would be a large expense for which there is little or no community support. The development of focal plane arrays, techniques for more efficient aperture illumination, and, for spectroscopy, more flexible spectrometers that can observe multiple lines simultaneously is the direction that must be taken. For example, a 3x3 focal plane array will increase telescope mapping speed by an order of magnitude, a significant advance. Experiments with multiple feeds on the Bonn 100m and the Greenbank (NRAO) 300m already look promising. Acousto–Optic Spectrometers are now in use at a number of observatories. The very wide frequency coverage of the units at Nobeyama, for example, permit important simultaneous studies of several molecular lines. On the millimeter systems, hybrid correlators that permit observations in several disconnected bands at once for multiple line research are also in productive operation.

One interesting scheme is the plan to equip the world's largest filled aperture radio telescope at Arecibo with an optical spherical corrector. The improvement in efficiency, beam shape, instantaneous frequency coverage and flexibility will beneficially affect many projects. For example, the capability for studying the distributions of neutral hydrogen in distant galaxies will be much improved, and the arrival of pulsar signals can be measured over a much wider frequency band at one time.

At millimeter wavelengths, the last decade has brought receiver noise improvements of an order of magnitude. The result has been an explosion in new information and many exciting discoveries. The receiver improvements have been especially valuable by enabling the studies of galactic structure and the interstellar medium of our own Milky Way to be extended to other galaxies. The breakthrough came with the development of cooled Shottky–diode mixers equipped with the low noise cooled FET amplifiers discussed above used as IF amplifiers. With the new cooled HEMT amplifiers in the IF, these devices should provide receiver temperatures of $\sim 50°K(DSB)$, with 80–100K at the telescope, in the 2.5–4mm atmospheric window. Receiver temperatures are about a factor of two higher in the 1.3mm window and 3–5 times higher in the 0.8 mm window. Mixers employing Superconductor–Insulator–Superconductor (SIS) junctions promise to yield the lowest noise temperatures. Laboratory tests show that they have mixer temperatures close the quantum limit and an order of magnitude lower than Shottky mixer temperatures. They also may be operated with conversion gain. However, in actual telescope installations, with noisy IF amplifiers and optics, they have either the same temperatures as Shottky systems or are better by no more than a factor of two. Because they also require superconducting temperatures, they are overall no more effective in remote telescope locations than the Shottky systems. More development work with these devices will soon bring us to the noise limit imposed by the atmosphere. In anticipation of this limit, work is already in progress on the invention of focal plane arrays for the single millimeter antennas. Developments in the 3mm window at the Five College Radio Astronomy Observatory and in the 1.3mm window at the 12m telescope at Kitt Peak will make possible, for example, the rapid mapping of many galaxies in CO, a most important advance.

At submillimeter wavelengths receivers are improving rapidly. Continuum bolometers are being used to study the dust emission from galaxies and regions of massive star formation in our own galaxy. Superheterodyne receivers are being built with steadily improving noise temperatures for the study of higher excitation rotational lines of interstellar molecules and atomic fine structure lines in interstellar clouds containing massive stars and shocks. These receivers are in use at the highest ground based telescopes at wavelengths of 300μ and longer and on the Kuiper Airborne Observatory at shorter wavelengths.

2.   Astronomy at the Shortest Radio Wavelengths.   Instrumental developments and observations at millimeter and submillimeter wavelengths have been increasing at an accelerated pace during the past decade. One important reason is that radio emission from thermal processes, those that are in approximate local thermodynamic equilibrium, is strongest at the shortest radio wavelengths and offers the best opportunity for sensitive high angular resolution observations of basically thermal phenomena. A second motivation

is that the spectrum of the interstellar medium is much richer in emission lines, particularly rotational transitions of the many complex interstellar molecules, and these lines are invaluable for the study of the chemical and physical state and evolution of the interstellar medium. A third factor is that it is possible to observe the emission from interstellar dust at wavelengths shorter than about 3mm. This emission is optically thin and readily interpreted. Measuring the dust continuum from galaxies at 1mm will probably become the standard technique for measuring the masses of most galaxies. A further reason for the development of many millimeter wave telescopes is that a small telescope of modest cost may have useful sensitivity and angular resolution for important science.

The level of interest in this part of the radio spectrum is shown by the considerable number of telescopes recently completed or in the planning stages. Two large instruments are the Nobeyama 45m in Japan and the IRAM 30m on the Pico Valeta in Spain. Figure 1 shows spectra of the two isotopes ^{28}SiC$_2$ and ^{29}SiC$_2$ detected with the 30m telescope in the atmosphere of the carbon star IRC+10 216 (Cernicharo *et al.*, 1986). In addition, a 13.7m telescope will be built in the Chaidanu Basin in China, and a 10.4m telescope has been completed at the Raman Research Institute in Bangalore. There are several smaller instruments: a 4m at Nagoya in Japan, a 3m from Cologne University installed in the Swiss Alps, the Bordeaux University 2.5m now located at the Plateau de Bure (the IRAM interferometer site), and a 2.5m telescope at Sao Paulo University in Brazil. For submillimeter wavelengths, two telescopes have just been completed on Mauna Kea, Hawaii: the Caltech 10.4m and the James Clerk Maxwell 15m. A 15m antenna which will be operated jointly by ESA and Onsala is being assembled at La Silla in Chile. Further, a consortium of the MPIFR in Bonn and the Steward Observatory in Arizona is planning to construct a 10m reflector on Mt. Graham in Arizona. These are all in addition to existing telescopes at Onsala, Sweden (20m), Metsahovi, Finland (13.7m), Amherst (FCRAO), USA (13.7m), Kitt Peak (NRAO), USA (12m), Macdonald (Texas), USA (5m), New York (Goddard), USA (1.5m), Sydney (CSIRO), Australia (4m), and New Jersey (BTL), USA (7m).

In addition, there are several arrays for millimeter wavelengths. In California, USA, there is an array of three 6m antennas at Hat Creek and an array of three 10m antennas at Owens Valley. Both are presently operating at 3mm, and both plan to extend operation to 1.3mm and hope to increase the number of antenna elements. Figure 2 is a map of the nucleus of the star burst galaxy M82 observed in the 1–0 line of CO with the Owens Valley interferometer (Lo *et al.*, 1987). At Nobeyama there is an array of five 10m dishes, and at the Plateau de Bure, the IRAM is completing an array of three 15m antennas.

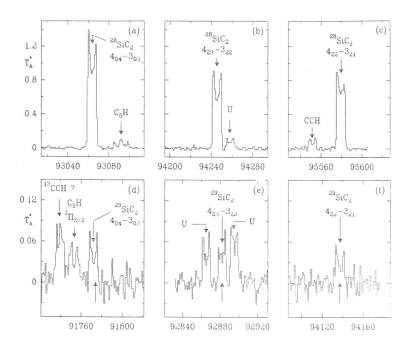

Fig. 1. Spectra observed toward the carbon star IRC+10 216 in three transitions of $^{28}SiC_2$, upper line, and three transitions of $^{29}SiC_2$, lower line. The observations were taken with the IRAM 30m telescope by Cernicharo *et al.* (1986).

A considerable amount of interesting science can be expected from these instruments. The smaller telescopes will map the large scale structure of the Milky Way and discover new chemical species. At the other extreme, the arrays will provide the detailed mapping of regions of star formation in our galaxy and the structure of nearby galaxies to red shifts of the order of 0.1. Figure 3 is a map of the galactic center molecular ring in HCN from Hat Creek (Gusten *et al.*,1987). The larger submillimeter telescopes will have good resolution by themselves, about 10" at best, and will be able to map low brightness emission in galaxies with considerable detail. A simple example illustrates the greater power of a millimeter telescope for studying

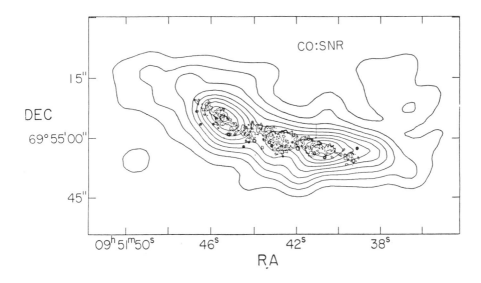

Fig. 2. An Owens Valley Interferometer CO(1–0) map of the nucleus of the starburst galaxy M82 (light contours) superposed on the VLA 6cm continuum which shows many bright supernova remnants near the CO peaks (Lo *et al.* (1987).

thermal emission relative to that of a centimeter telescope. If we compare the brightness temperature sensitivity of the five element Nobeyama array observing in the 2.6mm line of CO with the brightness temperature sensitivity of the 27 element VLA observing the 21 cm line of neutral hydrogen, we find that the former is more than two orders of magnitude more sensitive. The millimeter arrays currently have maximum baselines which give angular resolutions of about one arc second. With the receiver improvements of the past two or three years, the new brightness sensitivities will permit extensions for resolutions of a fraction of an arc second.

    3. Systems with Higher Angular Resolution. The most visible developments are the Very Long Baseline Array in the USA (VLBA), the European Very Long Baseline Network (EVN), and the Australia Telescope. As an adjunct to the VLBI systems, there are two space telescopes in the planning stage. One is the Soviet program RADIOASTRON which intends to put into

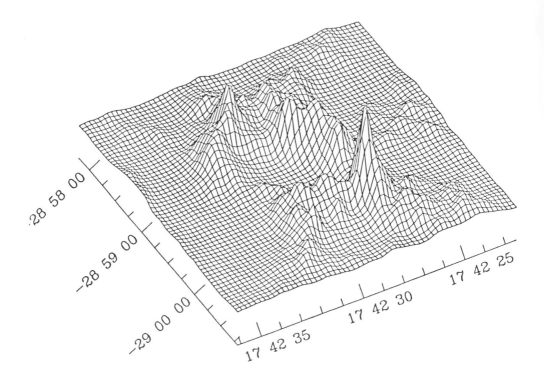

Fig. 3.   The molecular ring around the galactic center mapped in HCN by Gusten *et al.* (1987) with the Hat Creek Interferometer. The angular resolution is 3″. The height is the integrated line intensity.

orbit a 10m reflector by 1992, and there is the QUASAT project, joint between NASA and ESA, which plans to launch a similar sized antenna. There are also recent connected array developments at long wavelengths which obtain substantial increases in angular resolution.

The VLBA system, currently under construction, will place 10 new dedicated antennas into operation at sites across the USA capable of operation at wavelengths as short as 7mm, for the present, and possibly 3mm later (Kellermann and Thompson, 1985). Its expected completion is in 1992. The antennas are optimally located for good wavefront sampling and should provide much better calibrated data than is possible with the current *ad hoc* networks. The result will be maps with substantially improved dynamic range, better stability of registration on the sky, and a much improved short wavelength capability. When an antenna on a moving spacecraft is added to a VLB array, even better wavefront sampling results, and there will be a further improvement in the angular resolution with the additional longer baselines.

Unlike the millimeter wavelength telescopes which are best for the study of thermal processes, the VLB systems are intended for the study of basically non–LTE sources. For example, one observes the bright jet–like structures of radio galaxies and quasars and hopes to eventually understand the high energy sources in the galactic nuclei. One of the most spectacular achievements of VLBI has been the extension of the basic method of distance measurement by statistical parallax by two orders of magnitude in distance. In optical astrometry distances can be measured out to about 150 light years. Now the proper motions of bright spots of water vapor emission have given us a direct measure of the distance to the center of our galaxy, about 22,000 light years away (Reid *et al.*, 1986). With the VLBA it will be possible to extend such measurements to nearby galaxies and markedly improve our distance scale in the universe.

In addition to these cm wavelength VLBI experiments, there have been successful experiments both at very low frequencies, 82 MHz (Spinks *et al.*, 1986) and at high frequencies (Readhead *et al.*, 1983). Coherence was obtained in both cases, despite the ionosphere in the former case and the neutral atmosphere in the latter.

The Australia Telescope, expected to be completed for the centennial celebration, will have both a relatively compact connected element interferometer and an intermediate sized VLB array. Operation will eventually be at wavelengths as short as 3mm, permitting molecular line studies of the interstellar medium as well as high resolution observations of continuum sources. Both components of the system will add important high resolution capability in the Southern Hemisphere.

There is also a push toward higher angular resolution at the longest wavelengths. The 151 MHz synthesis telescope at Cambridge is presently producing maps at one arc minute resolution. Figure 4 is one field from the 151 MHz 6C survey (Baldwin et al,1985). A new low frequency array is planned for India, and there are new arrays in China and Chile. The Westerbork Synthesis Radio Telescope operates at 323 MHz, and the VLA is being equipped with receivers at this frequency. Ionosphere permitting, it will achieve an angular resolution of about 5″. These efforts will give us valuable pictures of radio

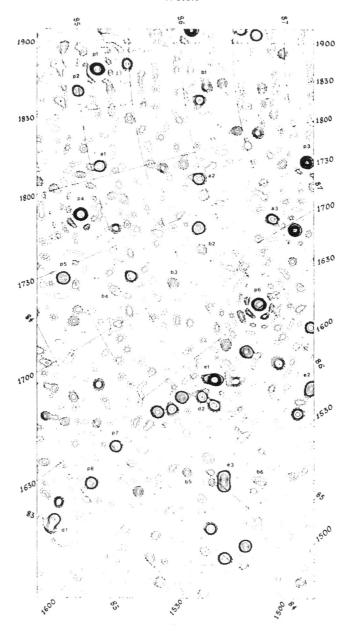

Fig. 4. One field from the Cambridge 6C survey at 151 MHz (Baldwin *et al.* (1985).

galaxies, and with the big increase in resolution that the VLA may achieve, there should be some unexpected results.

4.   Computer Developments.   It is a commonplace that computers have caused a revolution in all areas of science and technology, especially in those fields, such as astronomy, where numerical work has traditionally been extensive. Accurate and efficient handling of numerical data has indeed moved ahead rapidly. Computer control of telescopes is now common, with the result that the telescopes operate much more efficiently. The most important advance, over the whole of astronomy, has been in image processing. This area is particularly important for aperture synthesis arrays because of the way in which the observations proceed. Unlike conventional telescopes in which the detector or detector array is located in the focal plane, the array is effectively the entrance pupil of the telescope, and the data consist of the correlation function across the wavefront. The image is actually formed in the computer by means of the Fourier transform; the computer is an essential part of the telescope.

The required amount of computing to obtain maps from a given array data set has been increasing dramatically in recent years. There are several reasons for this. More and more frequently, observers are using multiple configurations of the big arrays, the WSRT and the VLA, to map larger fields with higher resolution. The antennas of the millimeter arrays have primary beams that are often smaller than objects that are mapped. Hence contiguous multiple fields must be observed and the resultant visibility data combined to produce the overall map, a field which then may contain many pixels. Second, additional correlators with variable delay have been added to enable spectral line mapping. This added dimension multiplies the number of maps. Third and most important, ever more sophisticated algorithms are in use to remove the sidelobes of the point spread function and to remove the errors caused by atmospheric phase fluctuations and imperfections in the antennas and receivers (Cornwell and Wilkinson, 1981). The latter procedure is known as "self–calibration".

Each of the tasks described above is already difficult to execute on the ordinary computers available to us, and we really must be able to handle them all simultaneously. For example, the spectral line part of the computation, an additional Fourier transform, should have its point spread function removed at the same time as it is done for the spatial part of the map. This should be carried out while multiple fields are observed and self–calibrated. For the big arrays, this program requires a supercomputer. Initial tests on large machines just with large field maps have produced remarkable results. Figure 5 is a 6cm wavelength VLA map of the jet in Cygnus A (Perley *et al.*, 1984).

There are now significant efforts to convert existing algorithms to run efficiently on supercomputers and to expand them to permit carrying out the large combined programs noted above. This development is the next big step in expanding the capabilities of the large centimeter arrays and the millimeter

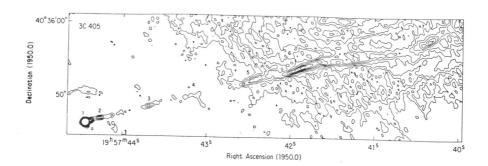

Fig. 5. A contour plot of Cygnus A at 6cm with 0.4″ resolution, showing the nucleus and the jet (Perley *et al.*, 1984)

arrays. There are important observing programs for which the data can be obtained but not be readily converted into maps at the present time. In addition to being able to execute a standard reduction routine, one must be able to test a variety of modeling schemes to compare the complex data sets with many different hypotheses, so that the bias in the reduction procedure is minimized. With the introduction of supercomputers it may also be possible to calculate realistic error estimates for the various reduction schemes. In the next few years we can look forward to important advances in the quality and usefulness in the large images from these efforts. Note that these ideas about image processing that are essential for the reduction of array data can also be applied to the maps produced by single dishes. It is customary to underilluminate single dishes in part to keep the sidelobes down. It would be more sensible to uniformly illuminate the dish, thereby increase its gain and resolution, and then remove the sidelobes in maps by the available image processing techniques.

THE NEXT HORIZON–SPACE

In the past two to three decades radio astronomers have been able to steadily increase the capabilities of their instruments by orders of magnitudes in the important parameters of receiver sensitivity, antenna collecting area, angular resolution, wavelength coverage, and spectral resolution. To continue this pace of development will require that we soon go into space with complex instruments. There are many important problems that can be attacked when this next step can be taken. Indeed, there have already been successful space radio astronomy experiments of modest scope, including the radio Astronomy Explorer Satellites (Weber et al, 1971) and the Voyager Missions to the outer planets (Warwick et al, 1977). More recently there was the successful use of a TDRSS satellite for VLBI (Levy *et al.*, 1986). The QUASAT and RADIOASTRON extensions to the present VLBI networks noted above are similar beginings of a large space radio astronomy program.

Moving our telescopes into space will permit observations over a larger wavelength range, with longer baselines, and in a lower noise environment. At long wavelengths, greater than 30m, the ionosphere becomes opaque, screening out the synchrotron emission from the most abundant of the cosmic ray electrons. At wavelengths shorter than about 0.3mm the absorption of atmospheric water vapor blocks our view of the warm dust and gas and embedded hot shocks of the interstellar media of our Milky Way and the galaxies beyond.

Weiler *et al.* (1985) have studied the potential properties of a very ambitious array of enormous extent and capability. This is the instrument that we might strive to build at some future time after smaller experiments, and it suggests what the future may bring. The array is comprised of thirty antennas of 50m diameter capable of operating over the wavelength range 10m to 1mm with a maximum baseline of 200,000 km. The emphasis of the scheme is very high angular resolution and sensitivity and micro–arc–second absolute position capability. Figure 6 shows the telescope's resolution compared with that of current and planned systems and the sizes of a variety of interesting astronomical objects. The doted line shows the expected resolution limit imposed by interstellar scattering. The resolution of maser clusters at large distances will extend parallax measurements to perhaps more than a million light years. At the highest frequencies, radio supernova outbursts may be mapped. Indeed, the structure of flares on stars can be mapped.

In the consideration of special purpose telescopes, an instrument for observations at the longest wavelengths deserves careful attention. Little is now known about the spatial distribution of cosmic noise at wavelengths longer than 30m. An array of a few dipoles whose orbits can be made to cover an adequate range of the visibility plane to synthesize a beam of the order of an arc minute could return valuable information from space. The detection of the turnovers in the spectra of synchrotron emission sources will provide measures of the magnetic fields. Then there are the steep spectrum sources identified with clusters of galaxies at slightly shorter wavelengths. Although the emission might be due to compton scattering of the microwave background radiation, the sources do not look like the cluster gas distributions. Since the electron lifetimes at these low frequencies are very long one might hope to observe fossil emission from galaxies whose central engines had gone out. At the longest wavelengths there might be coherent radiation observed from objects as bizarre as black holes. At slightly shorter wavelengths, near 10m, this array could map extragalactic radio sources with resolutions of a few arc seconds without the blurring effects of ionospheric refraction. Comparisons with the corresponding maps made with earth–bound cm wavelength

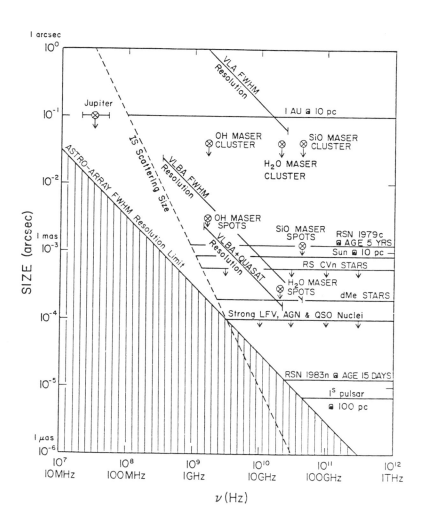

Fig. 6. Angular resolution versus frequency for a variety of instruments including the conceptual ASTRO–ARRAY of Weiler *et al.* (1987).

telescopes will provide considerable information on the distributions of both thermal and cosmic ray particle distributions. The fluctuation in the densities of thermal electrons in the interstellar medium can also be well studied with this instrument.

A space telescope to operate at submillimeter and far infrared wavelengths is of great interest at present. As noted above, the thermal component of the interstellar medium is best studied at these wavelengths. It is this part of the spectrum where the bulk of the dust emission is to be found, where the higher excitation lines of interstellar molecules in star formation regions and the fine structure lines of abundant atoms in shocks are strongest. For example, note that about one percent of the luminosity of the bright star burst galaxy M82 is emitted in the 61 micron line of oxygen. These strong signals will enable observations with the highest spatial and spectral resolution. The formation of stars and planetary systems can be observed with the needed resolution of 0.1″ or better. Several telescopes are planned, including FIRST, the ESA infrared telescope and SIRTF, the NASA infrared instrument. The most ambitious plan is the Large Deployable Telescope. Figure 7 is drawing of the present LDR concept, based on a large segmented primary mirror (Swanson, 1986).

A further step into space is the construction of telescopes on the moon, preferably on the far side. If there is ever to be a colony on the moon, it will be an excellent place for an observatory. The stability of the Lunar surface offers many advantages over free flying spacecraft for the construction of connected element arrays. Since there is no atmosphere, one can even imagine building an optical Michelson Interferometer array like the radio arrays that are now in use on the earth. The moon offers one further important advantage over space locations near the earth, relative freedom from interference. The radio interference background at the earth is steadily getting worse. For some projects it is a very serious problem. The observation of neutral hydrogen at large redshifts is becoming very difficult. Another program which requires an uncluttered spectrum is the radio search for intelligent life elsewhere in the universe. Serious searches are beginning, and will be badly hampered by the current level of radio interference. For these projects the back side of the moon will be the ultimate refuge.

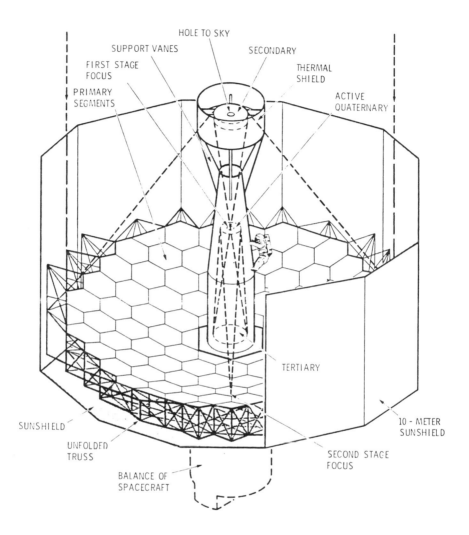

Fig. 7. A drawing of the present concept of the Large Deployable Reflector (Swanson, 1987). The primary mirror is 20m in diameter, composed of 84 hexagonal segments.

REFERENCES

Baldwin, J.E., Boysen, R.C., Hales, S.E.G., Jennings, J.E., Waggett, P.C., Warner, P.J., Wilson, D.M.A., (1985), The 6C Survey of Radio Sources — I. Declination Zone $\delta > +80°$, MNRAS, 217, 717–730.

Cernicharo, J., Kahane, C., Gomez–Gonzalez, J. and Guelin, M., (1986), Detection of $^{29}SiC_2$ and $^{30}SiC_2$ Toward IRS + 10216, Astron. Astrophys., 167, L–9–L12.

Cornwell, T.J. and Wilkinson, P.N., (1981), A New Method of Making Maps with Unstable Radio Interferometers, MNRAS, 196, 1067–1088.

Gusten, R., Genzel, R., Wright, M., Jaffee, D., Stutzki, J. and Harris, A. Aperture Synthesis Observations of the Circum–nuclear Ring in the Galactic Center, Ap. J., in press.

Kellermann, K.I. and Thompson, A.R., (1985), The Very Long Baseline Array, Science, 229, 123.

Levy, G.S. *et al*, (1986), Science, 234, 187.

Lo, K.Y., Cheung, K.W., Masson, C.R., Phillips, T.G., Scott, S.L. and Woody, D.P., (1987), Molecular Gas in the Starburst Nucleus of M82, Ap. J., 312, 574.

Perley, R.A., Dreher, J.W. and Cowan, J.J., (1984), The Jet and Filaments in Cygnus A, Ap. J., 285, L35.

Readhead, A.C.S. *et al.*, (1983), Very Long Baseline Interferometer at a Wavelength of of 3.4 mm, Nature, 303, 504, 1983.

Reid *et al.*, (1986), IAU Symp. No. 115: Star Formation, Reidel.

Spinks, M.J., Rees, W.G. and Duffett–Smith, P.J., (1986), Nature, 319.

Swanson, P. *et al.*, (1986), A System Concept for a Moderate Cost Large Deployable Reflector, Optical Engineering, 25, 9.

Warwick, J.W. *et al.* (1977) Planetary Radio Astronomy Experiments for Voyager Missions, Space Science Review, 21, 309.

Weber, R.R., Alexander, J.K., Stone, R.G., (1971), The Radio Astronomy Explorer Satellite, a Low Frequency Observatory, Radio Science, 6, 1085.

Weiler, K.W., Spencer, J.H, and Johnston, K.J., 1985, NRL Report 5657: ASTRO–ARRAY A Space-Based, Coherent Radio Interferometer Array.   Perley, R.A., Dreher, J.W. and Cowan, J.J., (1984), The Jet and Filaments in Cygnus A, Ap. J., 285, L35.

9

New communication networks

HELGA SEGUIN

ABSTRACT

With the important breakthrough of optical fibre techniques over the past few years a new type of communication networks is emerging offering a wide range of services. The techniques necessary for the development of these networks are available today but the difficulty consists in forecasting the appropriate service evolution. A long- term target solution and a pragmatic short-term approach are presented as two examples for the possible network design.

INTRODUCTION

Since several decades considerable investments have been made in order to establish performant and widespread communication networks. All countries tried to cover as quick as possible their whole territory with efficient communication tools connecting even the most sparsely populated areas. These high investments induced important technical progress in this field. The capacity and the flexibility of the systems increased continually while their cost and volume decreased. They became more and more intelligent offering a wide range of services to the user and the network operator. The operating and maintenance tasks have been as much as possible centralized and automated which enabled to achieve a fair profitability for the networks in spite of their high investment costs.

THE PRESENT NETWORKS

There are mainly two types of networks installed today :

. telecommunication networks with the telephone service as basic service ;
. broadcast networks for the broadcasting of television and radio programmes.

Both networks are installed and operated rather independently because their technical characteristics and the services offered to the users are quite different.

The telephone network has a rather limited bandwidth (about 3 kHz per telephone channel), but its architecture and its processing, control, signalling and switching functions are very complex. It must indeed enable to connect on demand any user with any other user of the network. The specific features of the telephone network are therefore : two-way switched links, star structure of the local network and interconnection of the telephone exchanges by a meshed urban and long-haul network. The transmission is done essentially on cable needing important infrastructures.

The broadcast network transmits broadband signals of about 6 MHz for TV channels and of about 15 kHz for sound channels. But, compared to the telephone network, its architecture is very simple and the main network function is a transmission function. The infrastructures needed are reduced but the network capacities are limited by the number of frequencies available within the spectrum allocated to broadcasting.

The two networks are today installed in all industrialized countries with penetration rates of about 100%. Most of them were installed in the sixties and seventies and the investments of today concern essentially replacement and modernisation of existing equipments.

With the liberalization of broadcasting and distribution of radio and TV programmes and the diversification of the telecommunication services, the present networks can no more satisfy the whole demand.

In the telecommunication field there exist a demand for the transmission of high speed digital and broadband channels and in the broadcast networks there is the problem of the limited number of available frequencies which leads to use more and more cable network solutions. In parallel with this evolution of the demand, there is the fact that the existing networks are almost amortized and often near the end of their lifetime and that the state of the art of the today's techniques can provide more efficient solutions. Indeed, the important breakthrough of optical fibre techniques over the past few years enables to conceive a new type of networks transmitting broadband and narrowband signals for distribution and interactive services. It is therefore tempting to install in the future multi-service and multi-client communication networks covering a wide range of services for business and residential use.

THE TARGET NETWORK AND THE DIFFICULTIES TO DEFINE IT

The multi-service network integrating a whole range of services is for the moment much more a concept than a reality. The existing networks have been designed and optimized for one special service, for example telephone or data or broadcast. The integrated service approach has been defined for the first time by the CCITT for the Narrowband Integrated Services Digital Network (NB-ISDN) giving to the subscriber access to two 64 kbit/s channels and one common signalling and data channel.

The same approach can be envisaged for a broadband network, that means each subscriber has simultaneous access to several network channels offering different services (e.g. access to TV and HiFi programmes, telephone, videophone, data services) which are controlled by a common signalling channel. In order to have a homogeneous network, it is assumed that all signals are digital and temporally multiplexed.

This is the target network actually discussed in international standardization groups and it is foreseen to be the basic communication network of the first decades of the next century. The envisaged network is entirely on optical fibres and wholly digital with bitrates on the subscriber lines of at least 140 Mbit/s.

The difficulty in defining such a future network is the fact that the whole service aspect is as good as unknown. New services, as for example the videophone are not at all defined. The bitrate actually discussed for such a service starts at 64 kbit/s and ends at 140 Mbit/s. The whole area of broadband retrieval services is as less known as all sort of multi-media services combining for example broadband and data channels or sound and data channels.

But without knowing the impact of these services on the network design, it is rather impossible to optimize this target network. And without knowing the demand and solvency of the services it is also difficult to prove its profitability. Therefore, the situation is actually the following : the techniques are available, and in hardware and in software, to design very powerful communication networks but the knowledge of the evolution of the services is not reliable enough to optimize these networks and to find the necessary investment funds which are considerable.

ONE WAY TO INTRODUCE NEW COMMUNICATION NETWORKS

In order to solve this chicken and egg problem of the future target network with respect to the new services which can guarantee its profitability, intermediate solutions must be adopted. One approach, chosen in particular in France, is based on a step by step approach.

The results obtained by several experimental local networks (e.g. BIARRITZ in France) and the fast progress registred in the area of optical long-haul links lead to the conclusion that the installation of an optical local network is now technically feasable. It is indeed no more difficult to install an optical cable infrastructure and its price becomes more and more competitive (figure 1). But what is still rather uncertain and also relatively expensive is the system built around the cable infrastructure that means the equipments for transmission, switching, signalling, processing, controlling and the different terminal equipments to be installed in the subscriber premises. The design of all these equipments is strongly dependent on the nature and the demand of the future services and is therefore difficult to optimize as long as they are not well known. That's why it seems interesting to explore

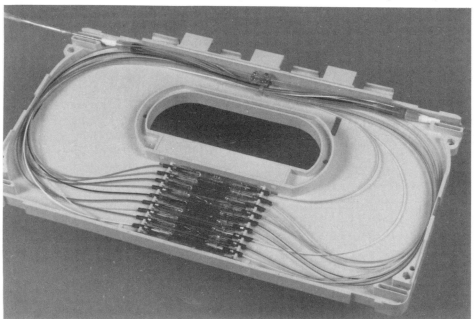

Figure 1 : Optical Fibre Splicing Unit

a solution which starts with the installation of a definite and comfortable, star-type optical cable infrastructure, but associate in a first phase a rather modest generation of network equipment which can be completed and modified later on according to the evolution of the service demand.

This approach seems to be a good compromise conciliating as much as possible different constraints : minimize initial invest- ments, assure nevertheless the future evolution of the network, without involving new expensive and long-term civil engineering work.

It does appear to be one of the most reasonable way to intro- duce rather quickly optical broadband networks which are in a price level comparable to traditional branch and tree coaxial networks but which offer potentially a much wider and quite different range of services.

The functional design of such a network is shown in figure 2.

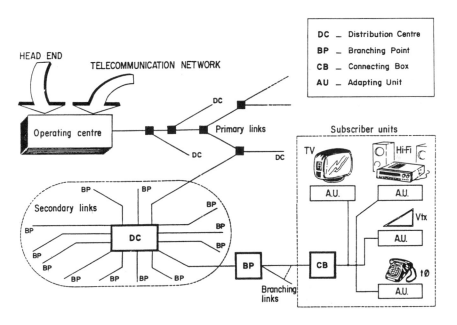

Figure 2 : Functional Design of a Broadband Multi-service
           local Network

An <u>operating centre</u> monitors and controls the network and assures the necessary adapting of signals between the broadband local network, the head-end and the telecommunication network.

A <u>primary link</u> network connects the different distribution centres to the operating centre. It includes one-way broadband links for distribution services and two-way digital links for interactive services.

The <u>distribution centres</u> are remote units which contain the terminal equipments of the primary links, the subscriber equipments for the channel-selection and the line interface and control units, in charge of the dialogue with the subscribers and the operating centre (figure 3). There are also included different processing units - necessary for the inter-connection of the telematic and professionnal links with the existing telecommunication networks. The number of subscribers connected to one distribution centre varies from some hundreds to about thousand.

<u>Figure 3</u> : Subscriber Broadband Switching Card of CD

The secondary and branching links connect the subscribers to their distribution centre.

Two subscriber units adapt the signals from the network into a form suitable for domestic TV and HIFI sets :

- . a connecting box is in charge of the optical-electrical conversion of the signals,
- . TV or HIFI adapting units, installed close to the receiver sets, provide the required input signals to the terminals and the controlling and monitoring facilities for the subscriber use (figure 4).

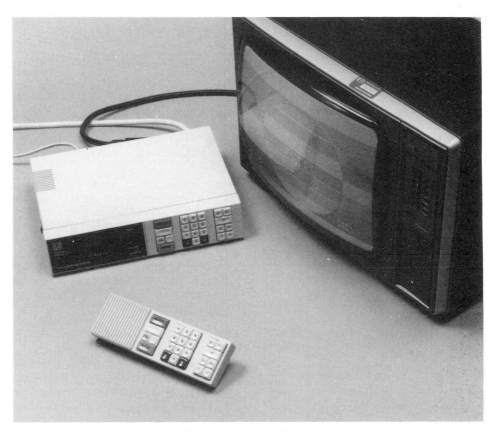

Figure 4 : TV Terminal Adapting Unit

The simultaneous services offered to the subscriber by the
first networks actually installed in France to the subscriber, are
the following :

- two TV channels, selectable out of 30 channels,
- one HIFI channel, selectable out of 30 channels,
- one two-way data channel of 4800 bit/s,
- one two-way NB-ISDN channel of 144 kbit/s.

This multi-service network is from the user point of view
rather comparable to a B-ISDN. The main differences with the future
target network are :

- distribution, data and telephone services have no common but
  separate signalling channels,
- the network is not wholly digital because the TV signal is
  transmitted in analog form,
- on the subscriber line there are only narrowband return
  channels and no broadband channel.

The reasons to install up from now such an intermediate not yet
standardized network are double :

- create an optical fibre broadband infrastructure,
- explore new services.

Both objectives will help to adopt later on a more performant
and more integrated network approach. Beacause even if we are defining today a target network, this will remain more or less a concept
which corresponds not always to the reality.

Communication networks are contunually evolving and include
techniques of several decades. This will be also the case in the
future broadband networks. That means it is much more important to
provide a flexible evolution of the networks then to stick only to a
target solution which will be difficult to reach.

CONCLUSIONS

The difficulties in introducing new broadband communication
networks are not technical ones. The main obstacles are the high
investiments necessary for the installation of such networks and the
rather unknown evolution of the service demand. Most countries are

today in a wait-and-see position and limit their efforts to experimental networks. But unfortunately, these rather limited networks give no significant results on the service aspects. That's why a pragmatic and evolutionary solution was adopted in France in order to try to explore the potential new services in parallel with the installation of a flexible network infrastructure. This seems to be one reasonable approach but others can be imagined. In any case there is no chance to introduce new communication networks if they are not justified by new service demands.

-o-

10

Digital optical techniques in computing and switching

J. E. MIDWINTER

ABSTRACT
Opto-electronic techniques are slowly penetrating into electronic processing systems, removing communications bottlenecks. In the future, they may have a more radical effect in spawning new processor architectures.

INTRODUCTION

In recent years there has been much discussion of "optical computers". Coming at a time when the advances in electronic processing machines have been and still are astonishing, one may reasonably ask what optics has to offer. To answer this question, we must first consider some of the background developments that have led to the present position since many of the ideas now espoused have undergone a long gestation.

In the early 1960's, much interest was aroused by the realisation that the diffraction of coherent light could be used to form the Fourier Transform of an object (see Preston 1972). Thus, the far-field radiation pattern of an antenna is the Fourier Transform of its nearfield pattern. In like manner, coherent light shone through a 2 dimensional mask carrying complex spatially modulated data yields the (spatial) Fourier Transform of that data in the optical far field, at the focal point of a suitably placed lens. In a single "clock" pulse, the whole data set can be Fourier analysed. Such processors exploit the free space propagation of light to carry out a very specific operation. Cross- and Auto-Correlation, Filtering and similar operations can also be performed. The problems of implementing such techniques tend to lie in the available dynamic range of the processing system, which is inherently analogue and limited, coupled with the problems of entering and extracting data from the system. Making and developing photos of the data is not attractive and greatly impedes the I/O operation! Attempts to overcome this using acousto- optic modulators etc have met with some limited success but do not threaten general purpose digital computation!

More recently, studies of the physics of non-linear optical interactions have shown that elements can be made that have optical I/O characteristics that might allow them to operate in a digital, binary logical manner and very simple logical circuits, such as ring oscillators, have been demonstrated, (Smith et al 1987). These have again revived interest in the use of light in processing systems.

Several other factors have also had an impact. Despite the impressive power of electronics, it is increasingly found that complex digital electronic circuits are being limited by the problems of precision timing and wideband communication at the chip level or between chips and boards. In parallel with this, the development of optical fibre communication systems has demonstrated beyond doubt the supreme interconnect capability of optics, (Midwinter 1983). It is natural to think of linking these observations. The possibility that some logical functions might be implemented in the optical domain may further assist this approach whilst the impressive power of lensed based imaging systems to provide parallel wideband interconnections through space is a further ingredient to be examined.

OPTICALLY ACTIVATED LOGIC
An enormous number of bistable optical elements have been reported in the literature, (Gibbs 1985). These are optical devices whose I/O reponse shows a marked hysteris in terms of optical power. Since they are power sensitive and "negative" power does not exist, the input/output curves are entirely contained within the positive quadrant where they show a hysteris response, starting at low transmission and switching to higher transmission at some thershold power as the input increases. Decreasing the input power leads to switch down at a lower power level.

The commonest form of such devices relies upon the use of a Fabry-Perot resonant cavity formed by two parallel mirrors between which is sandwiched a layer of non-linear material exhibiting an intensity dependent refractive index (or optical dielectric constant). The cavity is arranged to be tuned off-resonance in such a way that increasing incident optical power changes the refractive-index in the cavity to draw it nearer to resonance. This in turn enhances the internal optical intensity and can lead to an avalanche effect in which at some threshold power level, the device switches onto resonance. At the same time, the cavity switches from predominantly reflecting to predominantly transmitting. On reducing the optical incident power, switch down occurs at a lower optical power level, leading to a hysteresis loop in the I/O response.

The net optical output power is always less than the input power, since losses are always incurred. Hence to allow for "fan-out", it is necessary to provide an optical bias power that holds the device close to switching threshold and then to apply an additional signal power to trip it over threshold. With suitably accurate control of the bias and signal power levels, AND and OR gates can thus be simulated in transmission, NAND and NOR gates in reflection. However, to operate a large number of such elements in this way implies extremely precise control of optical power levels and the threshold level, (Wheatley 1987). Neither is readily achieved. The possibility of servo control being applied at the discrete gate level seems not to have been seriously contemplated. Since the switching threshold itself is normally a sensitive function of several variables, the devices incur the normal objections that go with threshold logic (Keyes 1985) plus the additional one that optical powers are notoriously difficult to control.

In operation, several other problems have to be faced. As the threshold, signal or bias power vary, so also will such operating variables as switching time, contrast and fan-out. It is perhaps not surprising to find, therefore, that despite a vast number of reports of single devices showing switching action and dating from at least a decade ago, the levels of "circuit complexity" so far achieved remain trivial at just a few gates. The absence of a "low impedance" optical output stage further restrains the flexibility open to the circuit designer.

Even a comparison of the reported results on discrete gates does not inspire much confidence (Midwinter 1985). Generally, we find that the devices so far reported switch relatively slowly, in the milli- to micro-second range and at power-speed products of nano- to micro-joules or worse. Some devices are projected to reach picojoule or just sub-picojoule switching energies with further development and thus promise comparability in this respect with electronic gates. The only region in which optical devices seem certain to operate free of electronic competition is in the sub-picosecond time range, where optical logic is ultimately expected to reach effective switching times of 10-100 femto-seconds. Such devices barely exist at present, are generally constructed in waveguide form, typically in optical fibres, and because of relatievly large power requirements are likely only to be used in simple logic circuits for ultra-fast multiplexing, demultiplexing, sampling or similar operations.

OPTICAL WIRING
Thus far we have identified almost no plausible reasons for seriously contemplating the use of optical technology in place of electronic technology in processing applications but have given many reasons for not using it. To find the arguments in its favour, we must return to the ideas in the Introduction and examine the interconnection aspects. These fall into several categories. In the first instance, we can postulate the use of optical "interconnects" to transport wideband data or timing information. This then leads us to consider the impact such technology might have on the electronic design process itself, which brings us to thoughts of architectural changes.

The first problem being attacked using optical interconnect is that of connecting printed circuit boards. Later, it is expected that the chip pin-out problem will be eased with optics. An optical signal can in principle be carried off chip at huge data rate via

an optical fibre. The engineering problem is to find a compatible
means of multi-/demulti-plexing the signal and converting it to and
from light. Typically, it is suggested that a linear array of LEDs
be hybrid or monolithically integrated along the side of a chip,
coupling light into a fibre array which carries the signal to a
matching array of photo-detectors. Since the optical transmission
medium is likely to offer negligible dispersion and a delay of
about 50 ps/cm, the data rate used will be dictated by the transmit
and receive components but might go up to 1Gbit/s. At present
100-200 Mbit/s is more accessible. A choice also has to be made
whether to use serial or parallel byte transmission; one fibre or
32 per connection! The technology for this is still being developed
and the problems of integrating the sources, detectors and fibres
intimately with packaged chips do not seem to have been fully
resolved. However, the approach does seem to promise major
increases in I/O data rates at chip level.

    The next step beyond this is to consider optical communication
within a single chip (Goodman 1984). Many possibilities are being
studied, ranging from optically distributing a single clock to many
separately located detectors within the chip, thus generating very
accurately timed local clock signals, through to transporting data
around the chip by optical means. Since optical beams can intersect
in space without cross talk, much interest centres upon the use of
some form of reflective coupling lying above and parallel to the
chip surface to channel signals from one part of the chip to
another. The reflector might well be a hologram and act not just as
a simple mirror but as a selectively directional mirror, directing
different source beams in different directions to discrete
detectors to provide many separate communications highways.

    Apart from the optical engineering problems of precisely
locating such an element relative to the chip, the technique is
also limited in application by the fact that suitable optical
sources are not readily monolithically integrated alongside the
electronic components. Many approaches to solving this problem are
being studied, ranging from using LEDs or surface emitting lasers
through to embedding electro-absorption modulators in the circuit
so that its status can be optically probed. Distributing clock or
data in this manner is expected to offer a solution to some of the
problems of clock skew and timing at high data rates. Ultimately,
it is not imposible to envisage clock or signal data being
delivered simultaneously to different parts of a large chip with
time precision measured in fractions of a picosecond.

    A simple lens performs an astonishingly powerful
interconnection in imaging one plane to another. Regarding the
plane as made up of a large number, say 1000x1000, resolvable
spots, each can be considered connected to its image point by an
almost infinite bandwidth, low cross talk "wire". Moreover the time
delay from each point in the object plane to its associated image
point is identical to very high precision, typically sub

picosecond. If we now think of our planar chip as the object, with emitters embedded in its surface, a lens offers a massively parallel wideband highway to some other suitably placed element. The highway is obviously bidirectional.

Bulk optic elements can also perform other functions on an "array" of data. Simple linear shifts or rotations are readily achieved on the whole data set (or image), albeit in "hard wired" form. The complete array of data can be "perfect shuffled" using a simple combination of lenses and prisms. (Lohmann 1985, Midwinter 1985). The optical system astigmatically magnifies the "object" by a factor of two along one plane only and the prisms shear that image to that the two images so formed are overlayed by half their area. The sequence achieved for a simple 8 number shuffle is thus:-

12345678

to

1 2 3 4 5 6 7 8

to

15263748

Again this is achieved on a whole array of data provided by the original object. Such a "wiring pattern" is a powerful ingredient in many processing algorithms since by succesive shuffle operations, it allows to the efficient association of data in all possible pairs within a finite data set. Such an operation is a key operation in sorting, switching, FFT calculation and some matrix operations. That it can be achieved across a large data set with zero time skew is of great interest.

From the above, it is clear that the designer's "wish list" includes a set of compoenents to allow him to switch as desired from the electronic to the optical domain at any convenient point within a complex circuit. This seems to imply the use of a III-V semicondutor technology, either for all the components are at least as an overlay on silicon for the optical output devices.

A particularly attractive I/O element is the Multiple Quantum Well Electro-Absorbtion Modulation (MQW.EAM) which can be used as a very fast photodetector and as a modulator to modulate wideband data onto an interrogation optical beam. These devices have been demonstrated to operate at data rates to 10Gbit/s and are thus fully compatible with the fastest III-V electronic logic elements Ref.10.

A DIGITAL-OPTIC SWITCH ARCHITECTURE

The sort algorithm using the perfect shuffle wiring format (Stone 1971) receives N numbers in random order and. by processing through a series of N/2 rows of exchange-bypass modules which compare nearest neighbour pairs and either pass them or exchange, finally deliver them to the N output ports in linearly ascending order. The decisons to exchange or bypass are made solely by comparison of the

RELATIVE size of the two numbers, always sending the largest to the same of the two ports at each point in the matrix. Such an algorithm provides the basis for a self routing switching matrix, (Midwinter 1987). The cross points must be "intelligent" exchange-bypass units that, following a reset signal, can read the next incoming "data" at the two input ports as address data, identify the largest address (a one-bit comparison operation) and latch the state to exchange or bypass for the following true data flow.

The logical layout for such a matrix follows that for thesort algorithm discussed above. In operation, all cross points must be reset, a set of the desired output port addresses injected into the matrix via the optical data paths and each row in turn of the matrix is then enabled to set and latch its own states. In terms of speed and complexity, we envisage the discrete cross points being formed in III-V electronic logic with optical I/O and clocking. All the key components already exist to achieve this so that a single cross point will probably consist of photo-detector inputs, electronic logic for the address comparison, latch and data cross point logic leading to electro-absorbtion modulator outputs into an optical "zero time skew" perfect shuffle interconnect between matrix rows. The use of optical clocking for each module should allow the use of clock rates of many Gbit/s, so that the potential exists for switching matrices of considerable complexity, say 64x64, operating at real time data rates of 1-10Gbit/s per port. The data flow through such a matrix, because of the impression of the address data on the data flow, would be in a pseudo packet format and it would require little alteration to operate directly on real time packet data.

The wiring for a 128x128 matrix implies over 5000 internal wideband connections, a forbidding number. However, a further advantage of this switching matrix structure is that the extreme regularity in its interconnection patterns, each being identical, promises a very simple solution to the problem. It lends itself to the use of a "free space" optical perfect shuffle interconnect. Take the matrix and chop it at the entrance to each row of logical exchange-bypass modules. Now stack these on top of each other. A two dimensional array of modules is thus obtained with all the interconnections in a form that can be addressed by a single lens system. A layout for making the connections of such a matrix has been described (Midwinter 1987) in which a reflective-perfect shuffle is used, linked with optical fibre array inputs and outputs and discrete optical clocks for each row of the matrix to allow for "ripple resets" throughout the matrix to ensure the minimum interruption of data flow during a reset operation.

Such a switch provides an ideal "test case" for the comparison of digital optic technology against electronic technology since the "switching computer" pkaces maximum emphasis on "interconnect" and minimum emphasis on "computing". Moreover, only in telecommunications is it taken for granted that wideband data will be transported in optical form. It is thus likely to be the first situation in which an interconnect advantage will be found in a system that in all other respects could be configured entirely in digital electronics.

ARCHITECTURAL IMPLICATIONS & COMPUTING

In the discussion above, we have concentrated on relieving communication bottlenecks within more or less conventional electronic systems. However, interest also centres on more revolutionary change. The potential for using a wideband parallel lens based highway to transfer very many bits of data in a single clock pulse implies the possibility of new forms of processor architecture, very different to that of todays Von-Neumann machine.

The best example is probably that of the optical implementation of a Hopfield Neural Network for pattern recognition (Psaltis 1985). Whilst the work so far remains at a fairly primitive level, the potential for development is clear. The operation is partially analogue and partially digital and inherently embodies a very high degree of parallelism in its implementation. The purpose of the "machine" demonstrated was to accept a 20 binary bit word and to match it, to best fit, to one of four stored words. The input 20 bit array consists of 20 optical sources, say LEDs. The 20x20 "memory mask" stores the known word data. If the words are signified by the integer i (1-4) and the bit values by j (1-20) then the words are $b(i,j)$. The transmission of the matrix $T(j,k)$ is arranged to be as follows:-

$$T(j,k) = b(i,j)*b(i,k)$$

$$T(j,j) = 0$$

If the input signal is then $S(j)$ ($j = 1-20$), and the optical system is arranged so that any one source illuminates uniformly row j or the $T(j,k)$ matrix and any detector k received the power from all elements in the kth column, then the output of the detector is given by

$$O(k) = T(j,k).S(j)$$

Such elements have long been studied for analogue optical multiplication. Adding an output element that thresholds the optical signal and returning it to the input to recycle it leads to system which very rapidly homes in on a stable best fit pattern at the output. Typically in this case, just a few cycles or clock pulses are required. The equivalent operation performed serially implies 400 multiplications and additions as well as a large number of accessions to RAM. Since the optical system clock pulse could quite readily be in the nano to micro second region, the speed gain inherent in this approach is instantly apparent and become greater as the problem is scaled.

In a practical implementation, it is almost certainly necessary
to work with many more and much larger vectors. However, volume
holograms are already well established as high density storage
media and lend themselves well to applications of this type.
Threshold detection elements with sufficient sensitivity and power
gain are still the subject of instensive study and the overall
system design for such a system whilst aspects such as the most
efficient data coding, analysis of noise performance etc have only
just attracted interest although the properties of such networks in
neural or electronic terms have been the subject of intensive study
for many years.

This system illustrates an interesting property of optical
wiring, namely that in addition to high levels of parallelism it
can also lead to very large effective connectivities. In the above
example, one could equally well allow light from every input
"pixel" to fall on every "memory" pixel. This might correspond to
an equivalent connectivity of order 1E6 or more. However, against
this, it is at present difficult to envisage an effective technique
for readily changing the "memory" (although materials with
optically addressable absorption or refractive index do exist in
photo-chromics and photo- refractives) so that a general purpose
processor again seems rather distant. Finally, we note that
inherent in these novel architectures there is implicitly assumed
the presence of a parallel space optical wiring component. Whether
this is solely coupled at the "logic" plane or "memory" plane with
purely optical elements or with hybrid combinations of optics and
electronics remains to be seen. These studies are still in their
infancy.

CONCLUSION
Starting from the observations that electronic procesing is often
limited by communication bottlenecks and optical transmission
systems by electronic processing, we have followed a path that has
led us to postulate complex hybrid opto-electronic switches or
processors. In few cases do there seem to be good arguments for
"all optical" implementations but the case for a greater
penetration of optics into processing systems is overwhelming, once
the ultimate in (electronic component) performance is sought. This
is limited at present by the lack of an engineered technology, even
though all the key components essentially exist. We can confidently
predict that the next five years will see many of these technology
problems being solved, thus opening the way to switches and
soecialist processors of unsurpassed performance.

REFERENCES

1. H M Gibbs, (1985) "Optical bistability, controlling light with
light", Pub. Acadamic Press, New York,

2. J W Goodman, F J Leonberger, S Y Kung & R A Athale, (1984) "Optical inter-connections for VLSI systems", Proc.IEEE, Vol.72, pp.850-866,

3. R W Keyes, (1985) "Optical logic in the light of computer technology", Optica Acta Vol.32, p.374

4 A Lohmann, W Stork, & G Stucke, (1985) "Optical implementation of the Perfect Shuffle", IEEE/OSA Topical Meeting on Optical Computing, Incline Village, Nevada, USA, 18-20 March,

5. J E Midwinter, (1984) "Optical fibre communications, present & future. Proc.Roy.Soc.Lond.A, Vol.A-392, p.247-277,

6. J E Midwinter, (1985) "Light electronics, myth or reality?", IEE Proceedings Vol.132, Pt.J, pp.371-383,

7. J E Midwinter, (1987) "A novel approach to the design of optically activated wideband switching matrices", Submitted Proc.IEE. Pt.J.

8. K Preston (1972), "Coherent optical computing", Mc.Graw-Hill, New York

9. D Psaltis & N Farhat, (1985) "Optical computing and the Hopfield model" IEEE/OSA Topical Meeting on Optical Computing, Incline Village, Nevada, USA, 18-20 March.

10. S D Smith, A C Walker, F A P Tooley & B S Wherrett, (1987) "The demonstration of restoring optical logic", Nature Vol.325, pp.27-31,

11. H S Stone, (1971) "Parallel processing with the Perfect Shuffle", IEEE Trans.Computing, Vol.C-20, pp.153-161,

12. P Wheatley & J E Midwinter (1987) (to be published)

13. T H Wood, C A Burrus, D A B Miller, D S Chemla, T C Daman, A C Gossard & W Weigmann, (1984) "High speed optical modulation with GaAs/GaAlAs MQWs in a PIN diode structure", Appl.Phys.Letters Vol.44, pp.16-18.

11

The encounters with Comet Halley, March 1986

W. I. AXFORD AND R. Z. SAGDEEV

INTRODUCTION

During March 1986, five spacecraft made observations of a comet,
Halley, for the first time. Two of these spacecraft, called Suisei
and Sakigaki, were launched by the Japanese space organisation ISAS
and passed several hundreds of thousands of kilometres from the
nucleus. Two spacecraft, VEGA 1 and 2, were launched by the Soviet
Union as part of the Interkosmos programme under the guidance of the
Space Research Institute IKI; these passed the nucleus at distances
of about 8000 km on 6 and 9 March. Finally, the European Space
Agency spacecraft Giotto passed at a distance of only 600 km from
the nucleus on 13 March. All encounters took place on the sunwards
side of the comet, which is the upstream side as far as the solar
wind is concerned.

The spacecraft carried a variety of experiments which performed
a wide range of in situ and remote sensing measurements and which
have provided a new level of understanding and of detailed knowledge
of the nature of comets (see Nature 321, 1986 and Proc. of the 20th
ESLAB Symp. on the Exploration of Halley's Comet, Heidelberg 1986).
It is not possible to give a complete account of all results and
their implications in a short paper but we will attempt to summarise
the main points. In fact, much of the data remains to be analysed in
detail so that further surprises may be in store.

A remarkable feature of the encounters with comet Halley was the
degree and success of international collaboration. This was particularly evident in the very effective use of the imaging experiments
on the VEGA spacecraft together with precise tracking and trajectory
calculations, largely provided by NASA through the Inter-Agency
Coordinating Group, IACG, which enabled Giotto to be aimed so precisely. International collaboration of a different nature was evident in the more than 1000 Earth-based observers from all disciplines and countries, who contributed to the International Halley
Watch.

THE NUCLEUS

The existence of a single monolithic nucleus of comet Halley was confirmed by the imaging experiments on VEGAs 1 and 2 and Giotto (Keller et al., 1986 and Sagdeev et al., 1986). It was found to be considerably larger than previously expected, being roughly ellipsoidal in form with principal dimensions of 16 km x 8 km x 8 km. The volume is ~5x10^{17} cm^3, corresponding to a mass of ~1-5x10^{11} tonnes. As is evident from the processed images shown in Fig. 1, the surface is largely inert with most of the gas (not visible) and dust being emitted in "jets" from perhaps only 10% of the surface. Some topography is apparent, notably a "mountain" several hundred metres high and a series of crater-like features with diameters of the order of a kilometre or so and perhaps 200 metres deep.

There is no evidence of ice on the surface, in complete contrast to some artist's impressions made earlier, and in general the albedo is very low (less than 5%), as required to make the observed dimensions consistent with the apparent magnitude of the coma-free comet seen at large distances from the Sun. The low albedo implies in turn that the surface temperature is quite high, and indeed the infrared spectrometer experiment on VEGA 1 indicated a colour temperature of 420 ± 60 K (Combes et al., 1986 and Emerich et al., 1986). This suggests that there is relatively little sublimation of icy material from most of the dark surface which must in some sense be rather stable.

Observations of the periodic growth and decay of neutral coma gases, notably the Lyman alpha observations from Sakigaki and ground-based observations of expanding CN shells, suggested that the rotation period of the nucleus is about 2.2 days, and this was apparently confirmed by the imaging experiments. However, subsequent interpretations of other ground-based observations suggested that the period is 7.4 days. It seems possible that this problem can be resolved if one takes into account the presence of several active regions on the surface of the comet and also the possible precession of the nucleus (Wilhelm, 1987).

NEUTRAL GAS COMA

The neutral gas comprising the coma of comet Halley was detected directly at a distance of the order of 2x10^6 km from the nucleus. Apart from an exponential decay with a scale length of several hundred thousand kilometres, the density distribution was inversely proportional to the square of the distance, indicating an approximately constant expansion velocity. Taking the latter to be 1 km/sec and assuming spherical symmetry, the total mass loss rate in the form of gas was approximately 10 tonnes/sec at the time of the encounter at 0.9 AU from the Sun (Gringauz et al., 1986), taking water to be the most important constituent.

HMC IMAGE #3475

MPAE – KRAMM 02/87 (33)

Fig.1 A closeup look of the nucleus of comet Halley taken by the Halley Multicolour Camera on Giotto. The Sun is on the left, 27° above the horizontal and 15° behind the image plane. The distance to the nucleus is 9,500 km, the resolution about 400 m. On the sunlit side round crater-like structures are clearly visible. (Copyright Max-Planck-Institut für Aeronomie.)

The neutral mass spectrometer on Giotto has provided the most detailed information concerning the structure and composition of the coma (Krankowsky et al, 1986; see Fig. 2). In particular it has confirmed that the dominant parent molecule is indeed H_2O and that there are also contributions from CO (~ 15%) and CO_2 (~ 3%). Isotopic ratios were also determined, notably for $^{18}O:^{16}O = 0.0023 \pm$ 0.0008 and D:H = $0.6\text{-}4.8\text{x}10^{-4}$, both of which are comparable to terrestrial ratios. This is consistent with the hypothesis that the oceans and atmosphere have a cometary origin. The D/H ratio is similar to that found in Uranus and Titan, but substantially larger than found in Jupiter and Saturn, suggesting that it is reasonable to consider the Uranus and Neptune systems as having been formed by the accretion of icy material in the form of comets with no large contribution from unfrozen H_2 and He (Eberhardt et al., 1986). Other isotopic ratios determined for Halley from the Giotto NMS experiment (Krankowsky et al., 1986) are $^{34}S:^{32}S = 0.045 \pm 0.010$ and $^{12}C:^{13}C =$ ~ 100 (with considerable uncertainty), together with the ratios found for magnesium isotopes in sporadic E layers (believed to be of cometary origin), indicating that there are no great differences from the Earth as far as the less volatile material is concerned.

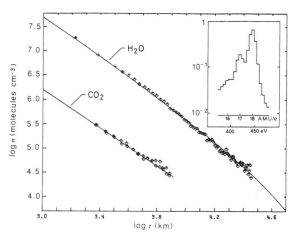

Fig.2 Water vapour (H_2O) and carbon dioxide (CO_2) number densities derived from the E-analyser data of the neutral mass spectrometer experiment on Giotto. (From Krankowsky et al., 1986.)

DUST

A range of cometary dust detectors was carried on the VEGA and Giotto spacecraft in addition to the cameras and other remote sensing experiments. The surprising results concerned the composition and the mass spectrum of the dust, which showed the presence of particles down to the smallest measurable mass, corresponding to about 10^6 atoms. The distribution of the dust was also a surprise, with small particles appearing at much larger distances from the nucleus than expected, given the effects of solar radiation pressure (Mazets et al., 1986; Simpson et al., 1986; McDonnell et al., 1986). Furthermore, there was evidence for some kind of clumping suggestive of the breakup of larger particles. The total mass flux from the comet

in the form of dust was comparable to, although perhaps less, than
that in the form of gas.

   The composition of the dust was determined in considerable
detail by time-of-flight mass spectrometry performed on the ion
cloud produced by direct impact on a known target (Kissel et al.,
1986a and Kissel et al., 1986b). The particles were found to be
quite heterogeneous with one important group being something like
chondritic, another dominated by magnesium, iron and silicon, and
most surprisingly a large group (especially among the smaller particles) comprised almost entirely of carbon, hydrogen, oxygen and
nitrogen (the so-called CHON particles). Examples of mass spectra
are shown in Fig. 3.

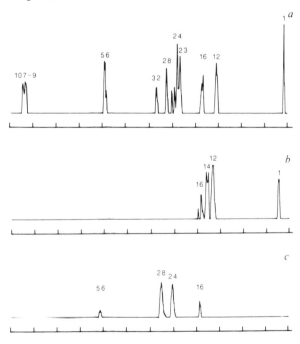

Fig.3 Time-of-flight spectra of three cometary particles measured
by PIA (internal reference numbers 33116 (a), 32778 (b) and 33675
(c)). Vertical scale is logarithmic; peaks are labelled with mass
numbers. (From Kissel et al., 1986b.)

   The CHON particles are evidently the most interesting in terms
of clues to the nature and origin of comets. They may be the result
of intense radiation processing at some early stage, either at the
time of formation of icy grains or later when the comet itself was
formed and then was exposed to the cosmic radiation for 4.5×10^9
years. In any case, their presence permits us to understand the
organic features in the infrared spectrum of the nucleus and dust,
the low albedo of the surface of the nucleus and the dust and the

apparent stability of most of the surface. The CHON material may be
sufficiently sticky to permit large lumps of otherwise low density
material to survive the sublimation of any embedded ices for a sufficiently long time for the smallest particles to be released at
large distances from the nucleus where they should not otherwise
exist. Furthermore, they may act as an extended source of molecules
such as CN and CO which seem to have parents which have not yet been
identified.

PLASMA

The general features of the interaction between the solar wind
and a comet have been fairly well understood for several years, however, there were some surprises, particularly close to the nucleus
where numerical simulations have poor resolution. On the largest
scale the most important phenomena are associated with the ionization of cometary neutrals and their subsequent pick-up by the solar
wind. This results in a slowing down and heating of the plasma which
can be detected at distances of several million kilometres and more.
The shock wave which is formed on the upstream side of the comet has
its Mach number reduced to no more than about 2, in contrast to 5 to
10 in the case of the Earth's magnetosphere, for example. Despite
its comparative weakness the shock was detected very clearly by the
plasma instrument on Suisei in particular (see Fig. 4).

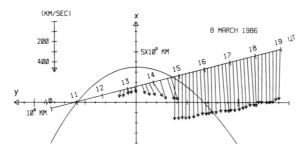

Fig.4 Plasma flow vectors obtained during Suisei's encounter with
comet Halley. The flow vectors and angles are represented in the
rest frame of the comet. (From Mukai et al., 1986.)

Throughout the region in which the solar wind penetrates the
cometary coma the pick-up ions, which are presumably mostly of the
water group, are very apparent in the form of a partially complete
shell in velocity space with the solar wind ions at the centre. This
distribution is unstable and the Alfvén/ion cyclotron mode turbulence which results serves to scatter the pick-up ions in pitch
angle as well as producing a certain degree of second order Fermi
acceleration to energies which can approach 1 MeV (Somogyi et al.,
1986 and McKenna-Lawlor et al., 1986). Close to the nucleus where
the solar wind speed decreases to quite low values due to the overwhelming effects of ion pick-up, the magnetic field becomes strong-

er, less turbulent and wrapped around the comet in the manner described many years ago by Alfvén. In this region the energetic pick-up ions are lost either by charge exchange and recombination or by diffusion and drift into space.

A remarkable feature of the plasma distribution near the comet was the "cometopause" found by the Plasmag experiment on VEGA 1 (see Fig. 5). At this boundary the slowly flowing external plasma was replaced by an essentially stationary plasma with heavy ions almost exclusively of cometary origin. These ions were found to be predominantly from the water group but with contributions also from CO, S, CO_2 and other more complex and heavier species. Very heavy ions with masses in the range 50-200 amu were detected by the PICCA experiment on Giotto and are possibly related to the dark material on the surface of the nucleus and to the CHON dust grains. It has been suggested that these are polymers, possibly polyoxymethylene formed by radiation damage in $H_2O/CO/CO_2$ ice mixtures at a very early phase of the formation of the nucleus (Huebner et al., 1987 and Vanysek and Wickramasinghe, 1975).

Very close to the nucleus (within 5000 km) the Giotto magnetometer found a very well-defined boundary within which the magnetic field dropped to zero (Neubauer et al., 1986) and the plasma became quite

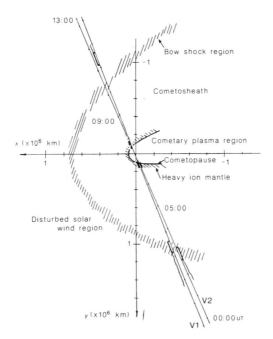

Fig.5a The plasma environment of comet Halley as observed by VEGA 1 and VEGA 2. Features of the PLASMAG-1 data are marked with symbols along the spacecraft trajectories (V1, V2).

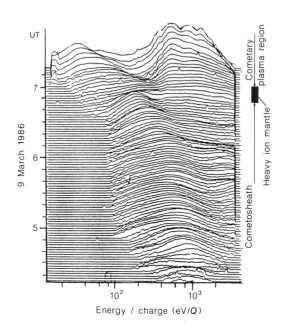

Fig.5b High-time-resolution ion energy spectra (2-min averages) measured by the VEGA 2 cometary ram analyser (CRA) during encounter.The crossing of the cometopause region is denoted by the filled rectangle. (Both Figs. from Gringauz et al., 1986.)

cold ($\leqslant$ 300 K) (Balsiger et al., 1986). This boundary is apparently sustained by friction between the outflowing neutral wind from the nucleus and plasma trapped in the magnetic field. The field itself showed reversals revealing that it had been captured from the solar wind over a period of at least many hours. In the case of the VEGA 1 flyby there was some evidence for magnetic field line reconnection which might be related to the tail disconnection events observed from the ground. There was no evidence of features which might be related to tail rays apart from some weak but clear periodic plasma density fluctuations observed by VEGA 1 (Gringauz et al., 1986).

CONCLUSIONS

Apart from providing some very detailed and in many cases surprising information concerning the environment of comet Halley, the results of the observations made from the five spacecraft have given us cause to think deeply on the origin of comets. Comet Halley may have made one or two thousand passes around the Sun and must have lost a substantial fraction of its original mass. It is therefore impossible to regard it as a pristine object in this sense, although the material which remains is presumably quite primitive. The dark organic material on the dust and on the surface of the nucleus play an important role in its evolution, although it is not clear whether it is largely confined to the surface or is prevalent throughout the body of the comet. These and other fundamental questions will be possibly answered as a result of future cometary missions involving penetrators and detailed analysis of subsurface samples. Until then, however, we can be well satisfied with the knowledge gained from these first survey missions.

REFERENCES

Balsiger, H. et al. (1986). Ion composition and dynamics at comet Halley. Nature 321, 330-334.

Combes, M. et al. (1986). Infrared sounding of comet Halley from Vega 1. Nature 321, 266-268.

Eberhardt, P. et al. (1986). On the CO and N_2 abundance in comet Halley. Proc. of the 20th ESLAB Symp. on the Exploration of Halley's Comet, Heidelberg (eds. B. Battrick, E.J. Rolfe and R. Reinhard), vol. I, pp. 383-386, ESA SP-250, ESTEC, Noordwijk.

Emerich, C. et al. (1986). Temperature and size of the nucleus of Halley's comet deduced from IKS infrared Vega-1 measurements. Proc. of the 20th ESLAB Symp. on the Exploration of Halley's Comet, Heidelberg (eds. B. Battrick, E.J. Rolfe and R. Reinhard) vol. II, pp. 381-384, ESA SP-250, ESTEC, Noordwijk.

Gringauz, K.I. et al. (1986). First in situ plasma and neutral gas measurements at comet Halley. Nature 321, 282-285.

Huebner, W.F. et al. (1987). Polyoxymethylene in comet Halley? Talk at the Symp. on the Diversity and Similarity of Comets, Brussels, 6-9 April.

Keller, H.U. et al. (1986). First Halley Multicolour Camera imaging results from Giotto. Nature 321, 320-326.

Kissel, J. et al. (1986a). Composition of comet Halley dust particles from Vega observations. Nature 321, 280-282.

Kissel, J. et al. (1986b). Composition of comet Halley dust particles from Giotto observations. Nature 321, 336-337.

Krankowsky, D. et al. (1986). In situ gas and ion measurements at comet Halley. Nature 321, 326-329.

Mazets, E.P. et al. (1986). Dust in comet Halley from Vega observations. Proc. of the 20th ESLAB Symp. on the Exploration of Halley's Comet, Heidelberg (eds. B. Battrick, E.J. Rolfe and R. Reinhard), vol. II, pp. 3-10, ESA SP-250, ESTEC, Noordwijk.

McDonnell, J.A.M. et al. (1986). Dust density and mass distribution near comet Halley from Giotto observations. Nature 321, 338-341.

McKenna-Lawlor, S. et al. (1986). Energetic ions in the environment of comet Halley. Nature 321, 347-349.

Mykai, T. et al. (1986). Plasma observation by Suisei of solar-wind interaction with comet Halley. Nature 321, 299-303.

Nature 321 (1986). 259-366.

Neubauer, F.M. et al. (1986). First results from the Giotto magnetometer experiment at comet Halley. Nature 321, 352-355.

Proc. of the 20th ESLAB Symp. on the Exploration of Halley's Comet, Heidelberg (1986). (Eds. B. Battrick, E.J. Rolfe and R. Reinhard), vols. I-II, ESA SP-250, ESTEC, Noordwijk.

Sagdeev, R.Z. et al. (1986). Television observations of comet Halley from Vega spacecraft. Nature 321, 262-266.

Simpson, J.A. et al. (1986). Halley's comet coma dust particle mass spectra, flux distributions and jet structures derived from measurements on Vega-1 and Vega-2 spacecraft. Proc. of the 20th ESLAB Symp. on the Exploration of Halley's Comet, Heidelberg (eds. B. Battrick, E.J. Rolfe and R. Reinhard), vol. II, pp. 11-16, ESA SP-250, ESTEC, Noordwijk.

Somogyi, A.J. et al. (1986). First observations of energetic particles near comet Halley. Nature 321, 285-288.

Vanysek, V. and Wickramasinghe, N.C. (1975). Formaldehyde polymers in comets. Astrophys. Space Sci. 33, L19-L28.

Wilhelm, K. (1987). Rotation and precession of comet Halley. Nature, in press.

SUBJECT INDEX

active experiments 97
adiabatic modes 21
adsorption spectroscopy 1
antennas 117
arrays 117
auroral oval 81

beam stacking method 21
bit-error rate (BER) 59
broadband ISDN 135
buildings 75

CHON-particles 155
clouds 75
code division 37
coherent detection 59
coherent optical communications 59
coherent time 1
comet Halley 155
comets 97
cometopause 155
cometosheath 155
cooperative queuing 37

distance and frequency
  standards 1

frequency-division multi-
  plexing (FDM) 59

gaseous adsorption 75
Gaussian beams 21
geospace 97
Giotto 155

heliosphere 97
heterodyne 59
high-frequency spectral
  techniques 21
homodyne 59

ice and snow 75
image processing 117
integrated services 137
interference holography 1
interferometer 117
ionosphere 81
ionospheric storms 81

laboratory experiments 97
land 75
laser 1
laser beam 1
low-noise receivers 117
lunar telescopes 117

magnetic storms 81
magnetosphere 97
millimeter wavelengths 117
modelling 81
multiple-access 37

network evolution 135
neutral gas coma 155
neutral wind 155
new communication networks 135
new services 135
nucleus (of comets) 155

optical communication 145
optical computing 145
optical resonator 1
opto-electronics 145

packet delay 37
photonic switching 145
plasma 155
plasma physics 97
plasma waves 97

Q switch 1
quantum noise 1

Index